Right House
Right Place
Right Time

Right House
Right Place
Right Time

Community and Lifestyle Preferences of the 45+ Housing Market

Margaret A. Wylde

BuilderBooks.com®
BOOKS THAT BUILD YOUR BUSINESS

A Service of
NAHB
NATIONAL ASSOCIATION
OF HOME BUILDERS

Right House, Right Place, Right Time:
Community and Lifestyle Preferences of the 45+ Housing Market

BuilderBooks, a service of the National Association of Home Builders

Courtenay S. Brown	Director, Book Publishing
Doris M. Tennyson	Senior Editor
Natalie C. Holmes	Book Editor
Torrie Singletary	Production Editor
Circle Graphics	Cover Design & Composition
McNaughton & Gunn	Printing

Gerald M. Howard	NAHB Executive Vice President and CEO
Mark Pursell	NAHB Senior Staff Vice President, Marketing & Sales Group
Lakisha Campbell	NAHB Staff Vice President, Publications & Affinity Programs

Disclaimer

This publication provides accurate information on the subject matter covered. The publisher is selling it with the understanding that the publisher is not providing legal, accounting, or other professional service. If you need legal advice or other expert assistance, obtain the services of a qualified professional experienced in the subject matter involved. Reference herein to any specific commercial products, process, or service by trade name, trademark, manufacturer, or otherwise does not necessarily constitute or imply its endorsement, recommendation, or favored status by the National Association of Home Builders. The views and opinions of the author expressed in this publication do not necessarily state or reflect those of the National Association of Home Builders, and they shall not be used to advertise or endorse a product.

Printed in the United States of America

12 11 10 09 08   1 2 3 4 5

ISBN-13: 978-0-86718-628-4
ISBN-10: 0-86718-628-3

Library of Congress Cataloging-in-Publication Data

Wylde, Margaret A.
  Right house, right place, right time : community and lifestyle preferences of the 45+ housing market / Margaret A. Wylde.
     p. cm.
  Includes bibliographical references and index.
  ISBN 978-0-86718-628-4
  1. Construction industry—United States. 2. Homeowners—United States—Statistics. 3. Housing development—United States—Statistics. 4. Dwellings—Design and construction—United States—Statistics. 5. Market surveys—United States. I. Title.
  HD9715.U52W95 2008
  690'.80688—dc22

                                                    2007041383

For further information, please contact:

BuilderBooks.com® NAHB
BOOKS THAT BUILD YOUR BUSINESS   A Service of NATIONAL ASSOCIATION of HOME BUILDERS

National Association of Home Builders
1201 15th Street, NW
Washington, DC 20005-2800
800-223-2665
www.BuilderBooks.com

Dedication

To all of the wonderful builders, developers, and trade associations that have allowed us to study the preferences and characteristics of their consumers; to all of the research study participants who have helped us learn what they want in their homes, neighborhoods, and communities; and to the builders who contributed their thoughts and ideas to this book.

Contents

Figures

Chapter 4. Home, Location, and Satisfaction
With Current Residence

Chapter 5. 45+ Renter Households

Chapter 6. Movers: Those Planning to Move and
Those Who Moved Recently

Chapter 7. Preferences for Location, Home Site, and Community

Chapter 8. Preferences for the Style and Appearance of a Community

Chapter 9. Community Amenities

Chapter 10. Home Preferences: Footprint, Floor Plan, Style

Chapter 11. Rooms and Spaces in the Home

Appendix A

Tables

Preface

Over the years my company has surveyed more than one million households about preferences for homes, communities, and lifestyles. The more research I do and the more that I listen to customers, the more I know that we have much more to learn from them. Fortunately, when builders and developers focus on a target market segment interested in their developments, they can learn just about everything they need to know, and fairly rapidly, to build the right house in the right place at the right time.

This book is based predominantly on a new large-scale research study that the ProMatura Group funded and conducted. We undertook the research out of frustration as we watched stereotypical age-qualified homes and communities being built and marketed as the product that boomers really want. As we watched communities falter and decades-old companies fail, we knew that builders and developers needed objective, current data to help them design the communities, neighborhoods, and homes that would really attract the burgeoning boomer market.

Birds of a feather flock together, but 45+ buyers are not the flock; rather, they comprise the entire class of all birds. Not all birds are alike. Boomers and silents include hundreds of market sectors. The better that builders and developers can identify and pinpoint the preferences, desires, and pocketbook of their particular target market, the greater their chances of building the right house, in the right place, at the right time.

Acknowledgments

I am grateful to BuilderBooks and the wonderful people in that organization for the opportunity to write this book. I am humbled by their talents, dedication, and unwavering spirit. In particular I want to thank Natalie Holmes who deftly edited this book, Doris Tennyson for her guidance and perseverance, and Courtenay Brown for her strong guiding presence. The team at ProMatura Group—particularly those who helped collect the data and prepare the information to be incorporated into this book—have my unending gratitude and my continued admiration for their professionalism, dedication, and talents. Thank you Bernie Smith, Edie Smith, Brenda Carothers, Heather Houghton, Bryon Cohron, Stephen Dziduch, Connie Hay, John Dee Warthman, Haley Downs, Katie Griffith, Matt Dickerson, Pam Massey, Glenn Downs, and the many talented research associates in our research call center for their thousands of hours of work on the study that helped build this book.

And finally, to my husband: Thank you for all that you do to let me be who I am.

About the Author

Margaret A. Wylde, Ph.D., heads ProMatura Group LLC, an international research firm headquartered in Oxford, Miss. Dr. Wylde has conducted research about home preferences and purchase decisions for more than 20 years. Unrelenting in her effort to stamp out ageism, she teaches builders and developers how to gauge the preferences of their target market sector before they decide what to build. Dr. Wylde is a popular speaker as well as author who has taken her message to audiences of builders and developers in the United States and abroad.

About 45+ Housing Consumers

Boomers and silents, Americans ages 45+, represent a huge home buying market. Therefore, builders, developers, and other industry professionals need to know more about what these consumers want in their homes, yards, neighborhoods, and communities. To help answer questions about the homes and lifestyles these potential buyers desire, this book provides results from a comprehensive survey of thousands of 45+ consumers, offering insights into where they live today, where they might move, and what they want in a new home.

Research on preferences of prospective buyers summarized in the following chapters points the way for builders, developers, and other industry professionals to develop the homes and communities that will appeal to this dynamic and growing market.

Vital Information for the Housing Industry

Right House, Right Place, Right Time: Community and Lifestyle Preferences of the 45+ Housing Market examines the current residences and future home preferences of people ages 45+. Some in this huge market sector will move to a new home; many will buy second and third homes; some will decide to rent; others plan to stay in their current residence.

This book identifies these buyers, renters, and stayers. It also describes where they live now, where they plan to live, and the types of communities and residences they want. Specifically, this book offers the following information about 45+ middle-American housing consumers:

- their demographic characteristics
- information about their current residences and communities
- what they would like to have in a new home and community
- factors that might influence their home purchase decisions

A Growing Market

Roughly 68 million Americans are at least 45 years old, and by 2012 that number will rise to nearly 75 million (table I). By 2011, half of all home owners will be age 50 or older (Paul Emrath, pers. comm.). Boomers, born between 1946 and 1964, are 42–60 years old. Members of the silent generation were born between 1926 and 1945 and are 61–80 years old. Finally, people in the GI generation were born between 1900 and 1925 and are at least 81 years old. Boomers are more likely to move than those in the silent generation, and those in the silent generation are more likely to move than those in the GI generation. As the baby boomers age, the number of home owners in the 55+, 65+, and 75+ age-groups will increase proportionately.

TABLE I Number of 45+ households in the United States: 2007 and 2012

Age-group	2007	2012
45 to 54	24,245,000	24,516,000
55 to 64	19,578,000	22,783,000
65 to 74	12,149,000	14,847,000
75+	12,217,000	12,718,000
Total	68,189,000	74,864,000

Source: Emrath, Paul, *Profile of the 50+ Housing Market, 50+ Demographics,* 50+ Housing Council, 2006.

45+ Homeownership on the Rise

The number of people 45+ who own homes continues to grow for several reasons:

- Many people in this age-group are first-time home buyers, including those who are 55–74.
- Homeownership is highest among people ages 55–74 (fig. I).

- An increasing proportion of householders 75+ years old are maintaining homeownership. These households are not debt-averse. To the contrary, they are more willing today to carry debt, and to do so later in life, than were previous generations (Harvard, 2006).
- Changes in marital status produce new households, and greater numbers of households are changing their marital status today compared with previous generations. An analysis of U.S. Census data revealed that 51% of U.S. women were living without a spouse in 2005, up from 35% in 1950 (Roberts, 2007).

Sources of Data on the 45+ Market

Most of the data summarized in this book is from a large-scale national research study completed by the ProMatura Group, LLC, between January and September 2006. ProMatura Group is a national research firm that has specialized in consumer housing research since 1984. More than 6,328 U.S. heads of households representing every state completed a comprehensive written survey about

- their current homes
- plans to move
- motivations for moving
- where they want to move
- types of homes and communities that they find appealing

This book focuses on the preferences and patterns of middle-Americans—people ages 45–64 with annual household incomes of $50,000 or more, and people 65+ years old with at least $30,000 in annual household income. Unless indicated otherwise, figures that accompany the text reflect only those survey respondents. Survey results were weighted to

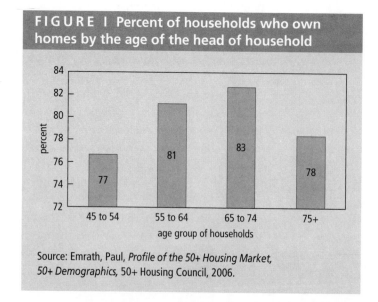

FIGURE I Percent of households who own homes by the age of the head of household

Source: Emrath, Paul, *Profile of the 50+ Housing Market, 50+ Demographics,* 50+ Housing Council, 2006.

accurately represent all U.S. households in these age- and income-groups, but they do not represent the entire U.S. population that is 45+. A larger proportion of the sample owns their homes, and the sample includes neither wealthy nor low-income households. Also, respondents from Alaska and Hawaii were eliminated from the final results because the sample sizes for those states were too small.

Segments of the 45+ Market

Survey data was collected and analyzed according to various subcategories that comprise the 45+ housing market, including the following:

- Age: 45–54, 55–64, 65–74, 75+ years
- Generation: boomer, silent, GI
- Life stage: obligation, transition, discretion
- Annual household income: <$30,000; $30,000–$49,999; $50,000–$69,999; $70,000–$99,999; $100,000–$149,999; $150,000+
- Home value: <$150,000; $150,000–$199,999; $200,000–$299,999; $300,000–$399,999; $400,000+
- Region of the country: 10 ZIP Code regions, U.S. Census regions
- Likelihood of moving: moved (in the past two years), move within 5 years, move in the future, never move
- Likelihood of moving to an active adult community or not: active adult, either, all-age
- Highest price movers expect to pay for a new home: <$150,000; $150,000–$199,999; $200,000–$299,999; $300,000–$399,999; $400,000+

Depending on the particular survey question or questions being discussed, a variety of additional categories discussed throughout the book help builders and developers define who the 45+ households are and where, when, why, and to what kind of community, neighborhood, and home these buyers will want to move.

Lessons for Builders and Developers

In addition to summarizing results from the national survey, *Right House, Right Place, Right Time: Community and Lifestyle Preferences of the 45+ Housing Market* includes lessons learned from more than 40 other site-specific studies completed by the ProMatura Group during a recent 24-month period. These studies examined consumers' perceptions and preferences for buying or renting all styles of homes in master-planned and conventional neighborhood communities with as few as 30 homes to developments that will someday have more than 30,000 homes. Where applicable, census data and information from other relevant national studies augment survey details about the market sectors.

Readers should keep in mind that heads of households are not distributed evenly by age and income in every city nationwide and that future customers will vary from market to market. Therefore, builders, developers, and other industry professionals who are planning new developments must understand the depth of the markets by both age and income in their targeted geographic areas. The final chapter of this book offers suggestions for doing so.

Organization of the Book

Using the survey and other data, this book offers a snapshot of housing consumers in the 45+ age-group and discusses their preferences according to a number of different dimensions, including

- household characteristics
- homeownership
- location
- moving and plans to move
- home plans and features
- types of communities
- community amenities

Part I discusses key demographics of housing consumers ages 45+, such as their particular life stage, marital and employment status, income, homeownership, and the amount they would spend for a new home. Part II looks at where these future home buyers live, whether they are owners or renters, how they feel about their current communities, and what might prompt them to move. Part III provides details about the types of homes, neighborhoods, and communities that buyers ages 45+ say are appealing, including floor plans, home features, locations, and amenities. It also offers market research tips and pitfalls to avoid when designing, building, and marketing homes for 45+ consumers.

Right House, Right Place, Right Time: Community and Lifestyle Preferences of the 45+ Housing Market provides builders, developers, and other industry professionals with vital information to create communities that will meet the needs of this demographic now and in the future.

WHO THEY ARE

45+ Middle-American Housing Market Sectors

The 45+ age-group is not monolithic. Rather it is a collection of market segments that vary not only by generation and income, but also by the types of homes and communities desired and the price people are willing to pay for a home. People in these markets may be single or married, first-time homebuyers or second-home buyers, working or retired, gardeners or golfers. This chapter defines the various sectors that differentiate this demographic and their home buying preferences and behaviors from one another. Information about these sectors and various subcategories within them are used throughout the book to help the reader make sense of the market data.

People in the vast age-group that comprises buyers 45+ years old may be just beginning a family or finding single life again after 60 years of marriage. Some of them are just reaching their stride in their careers, while others may be starting a new career. Many are ending paid employment altogether. At the same time, more and more are thinking about cutting back or changing the scope of what they do, but not quitting work entirely. Some retirees will be globetrotting; others will be immersed in civic duties; still others may be taking on new familial roles. All of these factors—stage of life, interests, and passions—will determine when, where, and to what type of home and community these buyers will gravitate, or whether or not they will move at all.

Nearly half of our customers today report they are still working, about 40% full time. It can really be a challenge to achieve a robust lifestyle around a working schedule. More attention needs to be paid to facilitation of resident connections and off-hours scheduling.

—DAVE SCHREINER
Vice President of Active Adult
Business Development
Pulte Homes, Inc.
Scottsdale, Arizona

Life Stages

Admittedly, housing choices and preferences differ among age-groups, but these differences may not be a function of people's age as much as they are a result of their particular life stage—obligation, transition, or discretion. For example, the 55–64 age-group is split nearly evenly between respondents who have obligations (44%) and those who have reached the discretionary stage of life (43%). Thirteen percent in the 55- to 64-year-old group are in the transition stage.

Among younger boomers, those in the 45- to 54-year-old age-group, almost 90% are in the obligation stage. During the obligation stage of life, people have two primary drivers: work and family. These two demands govern daily schedules; and decisions about where to live revolve around being close to work, providing a high-quality living environment, having access to good schools, and, if possible, being near other family members who can share responsibilities. Therefore, obligation stagers may not even consider moving until after children have left home for college, until responsibilities to elderly parents have been fulfilled, and/or until they change jobs.

The transition stage may last from a few months to several years. Transitions include

- planning to leave or leaving employment
- watching children leave home (or, perhaps, return home)
- becoming single again because of a divorce or the death of a spouse
- experiencing the death of parents

People in transition often are in the throes of making decisions about where they want to live and what they want to do as a result of this change in their daily lives.

Those in the discretion stage have moved beyond the obligation stage and have more freedom, flexibility, and choice in how they spend their time and lives. Some discretion stagers are adamant about doing what they want to do when they want to do it. Others may have traded a life of paid employment and care for children for one full of voluntary commitments to churches, causes, and grandchildren. Despite their full calendars, however, those in the discretion stage usually have greater control over what they do and when than their counterparts in the other two stages.

Household Income

Many heads of households may have ended their paid employment and therefore may be living on retirement savings and Social Security. However, although their incomes may be lower, their homes may have appreciated significantly and they may have investment assets in other real estate and retirement accounts.

As a rule, however, as income increases, so does the proportion of people that will move to a new home (fig. 1.1). Twenty-three percent of respondents with incomes of less than $30,000 are likely to move in the future compared with 37% or more of households with $100,000 or more in annual household income.

Still, household income by itself is not necessarily a predictor of whether or not that household will move in the future, particularly when one observes the likelihood of moving for individuals within the same age-groups, especially those 45–54. Within that group, those with the lowest incomes are as likely to move in the future as those with higher incomes. Also, householders ages 75+ with the highest incomes are the least likely to move of any group—by age or income.

Home Value Influence on Relocation

When home values are considered, across all age-groups, households with the lowest home values (less than $150,000) were the least likely to move. Still, a significantly greater proportion of 45- to 54-year-olds with the lowest home values are likely to move than people ages 75+ with the highest home values (more than $400,000).

Housing Preferences by Region

The region of the country in which householders live also influences the probability that they will move and their preferences for attributes of a new residence, neighborhood, and community. Therefore, responses for some survey questions are shown by ZIP Code region or the respondents' primary location of residency. A ZIP Code region map and table are included in Appendix A.

The lowest percentage of respondents who said they were likely to move in the future was in the Northeast (23%) and the Southeast (23%). The highest percentage was in the West, where 37% said they were likely to move in the future.

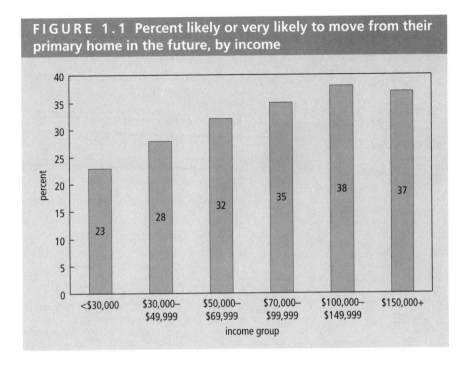

FIGURE 1.1 Percent likely or very likely to move from their primary home in the future, by income

By census region, heads of households in the Northeast ages 45+ were the least likely to move and, again, those in the West were the most likely. While 25% of the Northeasterners said they were likely to move in the future, 32% of Western U.S. residents said they were likely to move (fig. 1.2). A list of census regions appears in Appendix B.

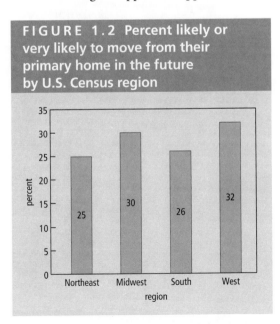

FIGURE 1.2 Percent likely or very likely to move from their primary home in the future by U.S. Census region

Active Adult or All-Age Community

The proportion of households that plans to move to an active adult community or that would consider moving to an active adult community has increased significantly in the past few years. Among households ages 45+, 22% are likely to move to an active adult community and 36% might move to an active adult community. Forty-two percent are likely to move to an all-age community (fig. 1.3). Many studies contrast those who prefer to move to an active adult community with households that prefer to move to an all-age community. These studies show that the householders who prefer to move to an active adult community are more active. The active adult householders are more likely to indicate that they participate in more recreational, educational, cultural, and sports activities and do so more frequently than their counterparts who prefer to move to all-age communities.

In contrast to an all-age community, in which homes are sold to people regardless of their age, an age-qualified active adult community is a community designed to provide housing exclu-

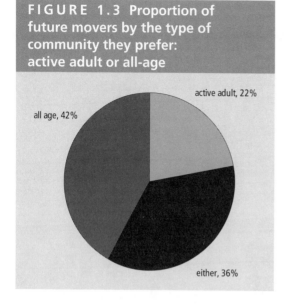

FIGURE 1.3 Proportion of future movers by the type of community they prefer: active adult or all-age

active adult, 22%

all age, 42%

either, 36%

sively for people 55+. These communities must comply with the Fair Housing Amendments Act to exclude younger households. The communities often provide any or all of the following:

- maintenance of community amenities and common spaces
- lawn and landscaping services
- exterior home maintenance

Other active adult communities may offer nothing more than an age-qualified community of homes. At the other end of the spectrum, some active adult communities have extensive amenities including the following:

- large and/or multiple community centers with outdoor and indoor swimming pools
- indoor walking tracks, golf simulation rooms, and bowling alleys

- ballrooms and theaters
- craft rooms for pursuits such as woodworking and electronics
- outdoor tennis courts, golf courses, and boccie ball courts

Housing Spending

The survey examined the preferences for homes and communities of 45+ middle-American households planning to move in the future. These households are categorized according to the highest amount they plan to spend on their homes. The largest percentage (28%) plans to spend $200,000–$299,999. The next largest group, 22%, plans to spend $400,000 or more. A breakdown including the remainder of households and their desired price ranges is shown in figure 1.4.

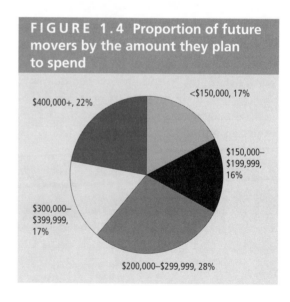

FIGURE 1.4 Proportion of future movers by the amount they plan to spend

$400,000+, 22%

<$150,000, 17%

$150,000–$199,999, 16%

$300,000–$399,999, 17%

$200,000–$299,999, 28%

2

Key Demographics of 45+ Middle-American Households

The survey included questions about marital status, employment, education, homeownership and home values, and size of 45+ households. In addition, heads of households were asked about "issues of housing as we get older," including whether respondents had any health impairments. Finally, respondents were asked about vehicle ownership, including recreational vehicles.

Marital Status

Although married-couple households became a minority of the total adult population in 2005, two-thirds of the survey respondents were married and another 3% lived with a domestic partner (U.S. Census, 2005a). One-third of the respondents were single. The number of single-person households will increase in the future because of changes in marital status and increased life expectancy. Longevity in general doesn't necessarily decrease the likelihood that one spouse will outlive the other. As married people age, they are more likely to be widowed. Only 2% of the 45–54 age-group is widowed, whereas 45% of people 75+ years old are widowed. With an ever-increasing proportion of single-person households, demand for home styles, sizes, and prices is likely to change.

A greater percentage of 45- to 54-year-olds than those who are ages 55+ are divorced, single and never married, or living in a domestic partnership. These statistics are consistent with two trends among adults in all age-groups today: marriage is being delayed, and marrying after becoming divorced or widowed is seemingly not as high of a priority as it used to be.

Currently, people who are married or widowed are more likely to own their homes than those who are divorced or separated, were never married, or are in a domestic partnership (fig. 2.1).

A change in marital status is significant not only because it changes living arrangements but also because it usually lowers a household's disposable income. In general, married-couple households have higher incomes and wealth than single-person households. When both spouses work, their combined income is even higher, and they can afford a higher priced house. In at least 72% of households with income of $100,000+, both spouses are working (Census 2005b).

Among heads of households in the 45+ age-group, 45% with annual incomes of less than $30,000 are married, compared with 88% of the group with incomes of $150,000 or more (fig. 2.2).

When looking toward future home sales, consider that although married heads of households have more income and wealth and can afford "more house," they also are less likely to be considering a move (fig. 2.3). At the same time, and contrary to what many believe, widows and widowers do not routinely move following a spouse's death. More specifically, widowers are slightly more likely to say they are never going to move compared with widows. The women are slightly more likely than the men to say they may move sometime in the future.

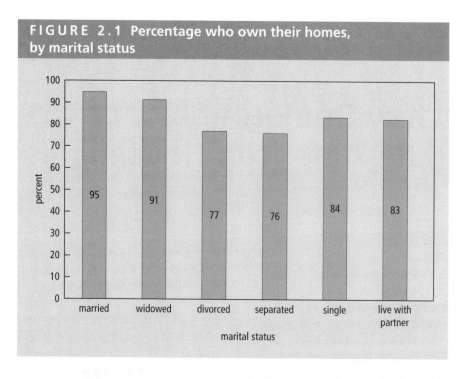

FIGURE 2.1 Percentage who own their homes, by marital status

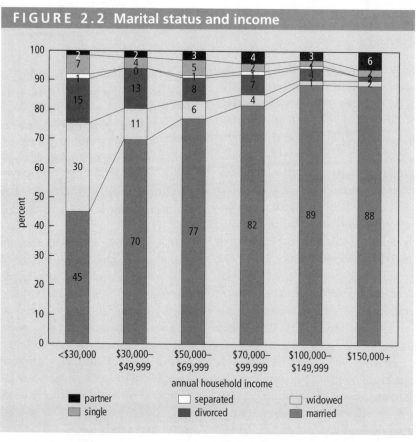

FIGURE 2.2 Marital status and income

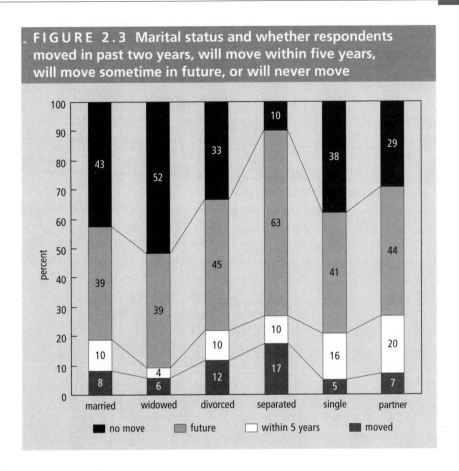

FIGURE 2.3 Marital status and whether respondents moved in past two years, will move within five years, will move sometime in future, or will never move

A range of home styles—including town-homes and condominiums—and price ranges may increase the opportunity to capture both single-person and married-couple home buyers, whatever their circumstances.

Education and Home Values

Homeownership is a common achievement for all groups, regardless of education or income. However, education, income, and wealth are inextricably tied, and these factors influence home buyers' interests and where they want to live.

The vast majority (96%) of 45+ middle-American heads of households have at least a high school diploma; 31% attended college; 24% have a bachelor's degree; and 15% have a graduate or professional degree.

Although education may be tied to wealth, it has not determined who will own a home; the data show that homeownership has been an achievable dream for all groups. Except for heads of households who completed only grade school, homeownership rates among middle-Americans who are 45+ years old top 90%, regardless of educational attainment. (For the former group, homeownership still tops 80%.) For most heads of households, particularly those in the lowest-income groups, the value of their homes is their largest single source of net worth.

Still, income and home values clearly correlate positively with level of education. No householders with less than a high school education reported an annual income higher than $99,999, whereas 37% of the householders with a graduate or professional degree reported their annual incomes exceeded $100,000. Moreover, 31% of 45+ middle-American heads of households with a graduate or professional degree own homes valued at $400,000 or more, but only 6% of householders with a high school diploma report the value of their home at $400,000 or more.

Markets in areas near universities and colleges and in other places in which adult educational attainment is comparatively high are likely to have more households with higher incomes.

Employment Status

Half of the survey respondents were retired, 29% were working full-time, 12% were working part-time, and 10% said they were not in the labor force. Because low-income heads of households were not part of the survey, the percentage of survey respondents who were working was less than for all 45+ U.S. households.

Considering the proportion of 45+ respondents who work at home (20%) and the 60% who own a computer (U.S. Census, 2005c), builders and other industry professionals can anticipate that more than half of these housing consumers might need space to work at home. Although most respondents who were employed full-time said they worked away from home (87%), 13% reported they worked from their homes (fig. 2.4). Among those who worked part-time, an increasing likelihood as people age, 37% worked from their homes. A combined 20% worked from their homes at least part of the time.

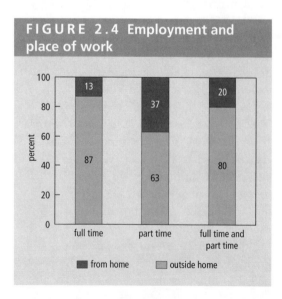

FIGURE 2.4 Employment and place of work

Similar to education, employment correlates positively with income. Seventy-six percent of the respondents who reported that their incomes were less than $30,000 also reported that they were not employed. In contrast, at least 63% of the heads of households with $70,000 or more in annual income were employed. Fully two-thirds of respondents with $150,000+ in annual income reported that they were employed full-time or part-time.

Among respondents who were employed, 57% said that they will fully retire eventually, and 21% said they will gradually phase out their paid employment until they are fully retired. The remaining working respondents said they would continue to work part-time forever (12%) or continue to work full-time (9%). The median age at which respondents said they would fully retire was 66.

Projections indicate that the average age at retirement will increase. Some heads of households will have to continue to work out of necessity, and others will continue to work because they can and they want to.

Although a majority of survey respondents said they would completely retire, the age of retirement is increasing for two primary reasons. First, the age at which a person can receive full retirement benefits from Social Security is increasing gradually from 65 for those born before 1937 to 67 for those born in 1960 or later. (A person can take partial benefits at any time after age 62.) Second, many individuals believe they will alternate periods of not working with periods of working. So, although many say they plan to retire or are retired, many will return to the workforce.

Recognizing these trends, developers of age-restricted housing are realizing that their menu of activities and events has to be more flexible to accommodate residents who are still working. It is likely, too, that the age range of households planning to move to active adult communities will expand in coming years because many who have opted to work longer have also opted to remain in their existing homes longer. Some plan to move to an active adult community when they quit working, which for a growing proportion will be beyond the age of 55.

Number Who Live in Home

Seventy-eight percent of middle-American heads of households in the 45+ age-group had two or more individuals living in their homes, and 22% said they lived alone. Fifty-five percent had two people in their households, 11% had three, and 12% had four or more people living in their homes.

About half (47%) of respondents ages 45–54 had three or more people in their homes. The proportion of households with three or more people living in the home dropped to 9% among householders ages 65–74. Nine percent of respondents ages 45–54 said they lived alone, compared with 63% of those who were 85 or older (fig. 2.5).

Less than half (43%) of survey respondents were couples that no longer had children living at home. Another 24% were single with no children living at home; 18% were couples with children at home; 10% were couples without children; and 4% were single householders with children at home.[1]

A majority of 55- to 64-year-olds become empty nesters. Only 7% of heads of households ages 65+ have children living in their homes. These people are likely to be sharing a residence with their adult children and/or grandchildren (Simmons, 2003). However, few residents of large active adult communities have their grandchildren living with them.

Universal Design

With people living longer, all home buyers are likely to benefit at some point from universal design. Therefore, homes can and should be de-

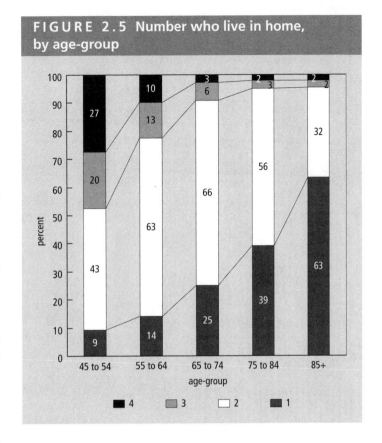

FIGURE 2.5 Number who live in home, by age-group

signed so that they are easier and safer to live in. Universal design ensures the usability of a product or environment by people of all sizes, ages, and abilities. Many difficulties people encounter in their homes are man-made. Yes, some people may be short, have difficulty raising their arms above their shoulders, or suffer pain when climbing stairs, but a home's design is what can make reaching kitchen cabinets difficult or climbing stairs necessary and painful. Although 75% of respondents 45+ years old said they had no health limitations or impairments, 30% of respondents 75+ years old said they had health limitations.

Among all U.S. householders 45+ years old, limitations in ability are linked with income for two primary reasons (fig. 2.6). First, people with higher incomes have enjoyed better living environments, nutrition, education, and health care. Second, households with lower incomes are more likely to be headed by a person who is

---

[1]Because percentages were rounded, they do not total 100%.

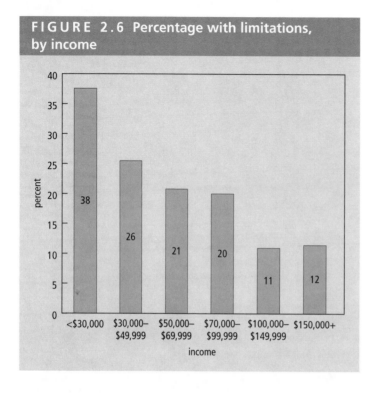

FIGURE 2.6 Percentage with limitations, by income

a motorcycle; and 1% had a snowmobile.

Six percent of survey respondents said they had "other" recreational vehicles or large items that they needed to store at their homes. Most of these heads of households said that they had more than one item. These items included airplanes (6 respondents); bicycles; canoes or kayaks; antique or classic cars; heavy equipment, such as backhoes and other construction equipment; golf carts and utility or other trailers; jet skis; motor homes; mowers and tractors (including lawn and farm tractors); other types of lawn and landscaping equipment (such as leaf blowers and snow blowers); scooters and wheelchairs; and woodworking equipment.

Households in the Western Mountain region (26%), on the Northern Plains (18%), and on the Pacific Coast (17%) were significantly more likely to own recreational vehicles and other equipment than those in the Eastern and Southern United States.

75+ years old. Thirty-eight percent of respondents reporting less than $30,000 in annual income said they were limited because of an impairment or a health problem. In contrast, 12% of respondents with $150,000+ in annual income reported having health limitations.

RVs, Boats, and Other Big Toys

Twelve percent of the respondents own a recreational vehicle, but most of these people are 45–74 years old. Only 2% of those who are 85 or older own one. People with incomes of $30,000–$50,000 are more likely to have RVs or other large items stored at their homes than are households with higher or lower incomes. Also, householders with home values of less than $150,000 to $299,999 were significantly more likely than those with the highest home values to have large recreational vehicles or equipment on their property. Of those who owned recreational vehicles, the greatest proportion owned a boat (12.2%); 6.1% owned a four-wheeler; 5.8% had

Redefining Midlife and Beyond

One thing is clear about middle-Americans in the 45+ age-group. Their needs are diverse and ever-changing. Therefore, builders and developers can expect that changes in marital status, household composition, employment status, and other characteristics that make baby boomers unique will continue to redefine the way they live.

Change in one's life often means changing where to live. Relocation may not be immediate, but often within a few years of a change—watching children leave home, changing a job or retiring, divorcing or becoming widowed—homeowners in this age-group will move.

Homeownership Among 45+ Middle-American Heads of Households

3

Americans are home consumers, and once they purchase a home, they continue to be homeowners well into their 80s. On average, buyers 45+ years old have owned more than two homes in their lifetimes, with the number of homes owned increasing with age and income. With extensive experience as home owners, these buyers are likely to be discerning customers: concerned about design, quality construction, and location. Although a house is first and foremost a home, it is also a long-term investment, and these buyers are likely to view it as such.

Builders should be prepared to reassure buyers that the home they are choosing is also a wise investment because without the knowledge and experience to know what will sell, buyers sometimes make mistakes. For example, a few years ago, I watched a couple build a house with several small rooms and no open space. The couple eventually moved to a larger home directly across the street because they felt claustrophobic in the home they had built. Although all of the other homes sold rapidly in this popular development, their former home remained unsold.

Characteristics of Home Owners

Homeownership is at its highest among 50- to 74-year-olds; between 75% and 83% of heads of households in this age range own their homes.

Homeownership increases with household wealth and income. Still, the vast majority of people in all income groups own a home, and half of respondents with less than $30,000 in annual income owned their home outright (without a mortgage payment). All but 3% of 45+ middle-American heads of households with incomes of $150,000 said they owned their homes, and 86% of those with less than $30,000 in annual household income owned their homes.

The longer heads of households have lived in their homes, the greater the probability that they own them. Ninety-nine percent of householders who had lived in their homes for more than 30 years owned them. In contrast, 72% of those who had lived in their homes for less than 2 years owned them. The longer people lived in their homes, the less likely they were to move. Therefore, builders planning communities in areas with many older neighborhoods may find selling to long-term residents difficult.

Home Value and Equity

Seventeen percent of respondents said the value of their homes was less than $100,000; 31% reported their home's value at between $100,000 and $199,999, and 51% said their home was worth at least $200,000. Thirty-two percent of householders ages 55–64 own homes valued at $300,000 or more. Figure 3.1 illustrates that 24% of this age-group has $300,000 or more

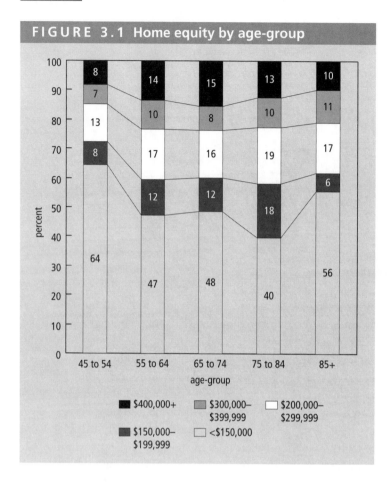

FIGURE 3.1 Home equity by age-group

Legend:
- $400,000+
- $300,000–$399,999
- $200,000–$299,999
- $150,000–$199,999
- <$150,000

The amount of home equity varied by home value and generation of the respondent. GI-generation heads of households whose home values were estimated to be $400,000 or more reported an average equity in their homes of $644,049.

With regard to a new home purchase, within each subgroup, the largest proportion of home owners said they would be willing to pay an amount in the range of their current home's value for a new home. However, only among home owners with homes valued at $400,000+ or more did a majority (65%) say they were willing to spend in that range for a new home. Figure 3.4 shows the amount that home owners said they would be willing to pay for a new home, based on the value of their current home.

in home equity. Between 40% and 64% of all 45+ households have less than $150,000 in home equity.

Home value increases with income, but values vary significantly by ZIP Code region (fig. 3.2).

Half of the heads of households who reported having an annual income of less than $30,000 said their home's value was less than $150,000. More than half (56%) of heads of households who reported annual incomes of more than $150,000 said their home's value exceeded $400,000. Homes in the Pacific Coast region had the highest values, whereas homes in the Midwest, Texas, and Louisiana (ZIP Code Regions 4–7) had the lowest. Figure 3.3 compares home value with home equity.

Boomers want it their way. Don't underestimate the wealth or appetite of the 55+ market, and don't assume baby boomers downgrade the size of their homes. They are willing to pay for generous spaces, great amenities, and conveniences. They want to entertain in their homes, and even if they don't use them much, they still like large kitchens.

—**ED HORD, AIA**
Senior Principal
hord/coplan/macht
Baltimore, Maryland

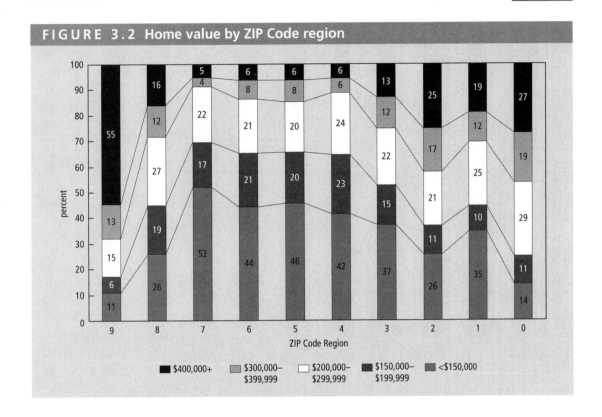

FIGURE 3.2 Home value by ZIP Code region

Legend: $400,000+ | $300,000–$399,999 | $200,000–$299,999 | $150,000–$199,999 | <$150,000

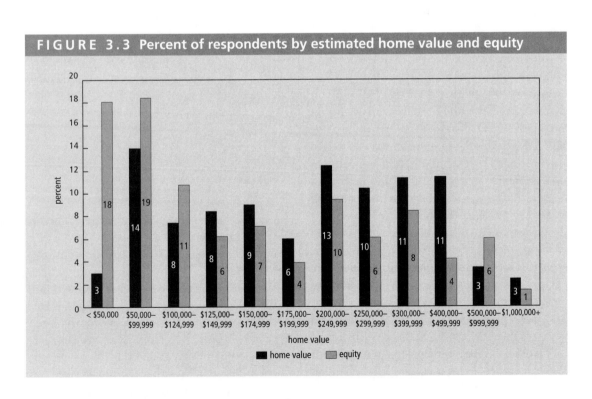

FIGURE 3.3 Percent of respondents by estimated home value and equity

Legend: home value | equity

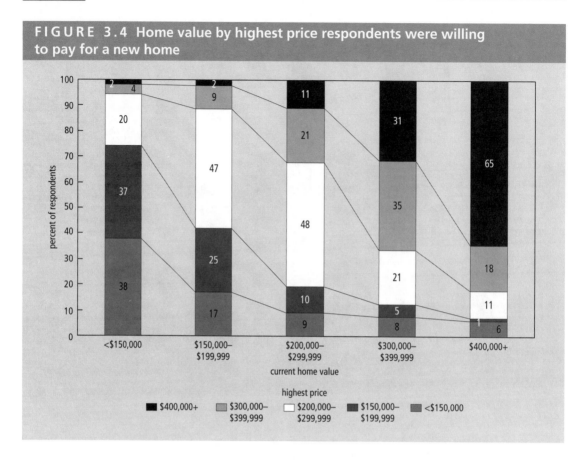

FIGURE 3.4 Home value by highest price respondents were willing to pay for a new home

Still, more than half of the middle-American home owners whose homes were valued at less than $200,000 said they wanted to pay more for their new homes than the value of their current homes. But within this group, the higher the home value, the greater the proportion that wanted to pay less for their new home than the value of their current home. Among respondents with homes valued above $200,000, less than one-third wanted to pay more for their new homes than the value of their current homes.

Heads of households planning to purchase homes in age-qualified communities were more likely to say they would spend $300,000 or more for their homes than were those who planned to purchase homes in all-age communities or who were unsure of the type of community in which they would purchase homes. Thirty-five percent of 55+ middle-American respondents planning to move to age-qualified communities said they were willing to pay $300,000 or more, compared with 28% of respondents planning to buy in an all-age community.

Mortgages of 45+ Middle-American Householders

Forty-four percent of middle-American heads of households ages 45+ said they owned their homes free and clear and did not have any house payment; 36% had traditional mortgages with fixed rates; 19% reported some other type of mortgage or arrangement; and 8% were renters. Only 1% of respondents had interest-only mortgages, a relatively new phenomenon. The proportion of home owners who owned their homes free and clear generally increased with age and the proportion who owned their homes with a traditional fixed-rate mortgage decreased with age (fig. 3.5).

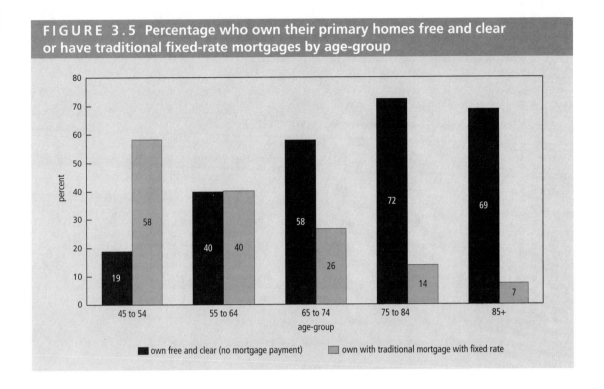

FIGURE 3.5 Percentage who own their primary homes free and clear or have traditional fixed-rate mortgages by age-group

The proportion of householders with variable rate mortgages decreases as age increases, but approximately the same percentage (3%–4%) of each 45+ age-group through age 84 carried home equity loans. Reverse mortgages first emerge among heads of households 65–74, and the greatest percentage (2%) with these mortgages was 85+ heads of households.

Heads of households with lower incomes are more likely to own their homes free and clear (fig. 3.6). The highest proportion of heads of households ages 45–64 who owned their homes free and clear was among those who reported the lowest household incomes. In general, although some home owners ages 45+ may have a large amount of home equity, they are likely to be older, with lower incomes, and not inclined to move.

Among 55- to 64-year-olds, the proportion that owned their homes free and clear decreased with increasing income, whereas nearly equal proportions of 65- to 74-year-old heads of households in all income groups surveyed owned their homes free and clear.

A greater proportion of householders who owned their homes with a variable rate loan had incomes of $50,000 or more (fig. 3.7). The greatest proportion of householders with interest-only loans had incomes of $150,000 or more, and the largest proportion with a home equity loan or a reverse mortgage had annual incomes of less than $50,000.

A slightly greater proportion of heads of households with homes valued at less than $300,000 reported owning their homes free and clear compared with heads of households who estimated the value of their homes at $300,000 or more. Fifty percent of the heads of households who reported their home values at $300,000 or more said they owned their homes with traditional fixed-rate mortgages.

Home owners with homes valued at $400,000 or more were significantly more likely to have variable-rate or interest-only loans than owners with homes valued at less than $400,000. The proportion with home equity loans decreased as home value increased. The greatest proportion of

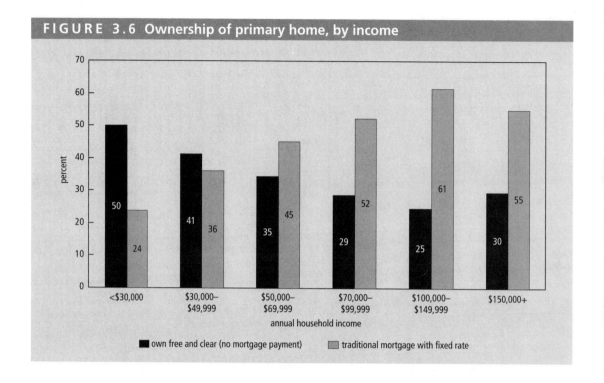

FIGURE 3.6 Ownership of primary home, by income

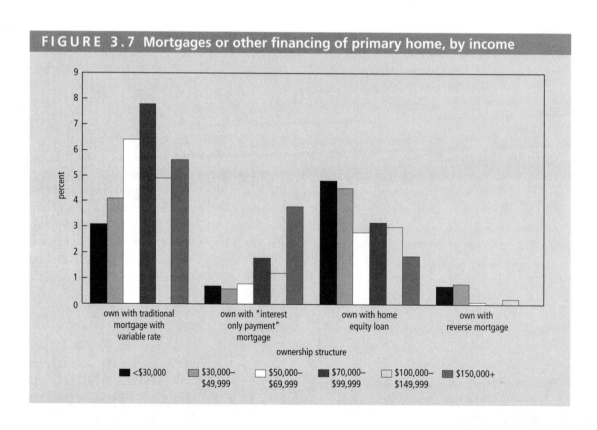

FIGURE 3.7 Mortgages or other financing of primary home, by income

respondents who said they had reverse mortgages had homes valued at $200,000–$299,999. Still, among all home owners, those with variable rate, interest-only, or home equity loans are a minority.

Ninety percent or more of all age and income groups reported that they owned their homes fee simple, although respondents 75 and older were more likely to have

- alternate mortgage structures such as reverse mortgages
- moved to a condominium
- moved to an entrance-fee community

Although the proportion of respondents that owned their homes as part of a condominium increased with age, only about 6% of heads of households ages 85+ said they owned a condominium. For all other 45+ age-groups the percentage was even less.

Householders who reported an "other" form of ownership fell into one of five groups:

- They owned their homes (both stick-built and/or manufactured) but leased the land.
- They neither owned nor rented homes because they lived with family members.
- Corporations owned their homes.
- They lived in parsonages.
- They had equity loans or reverse mortgages on their homes.

In describing who owned their homes, householders with the lowest incomes were the most likely to report "other" and that they were "living with a family member." None of the householders with either less than $30,000 or more than $400,000 in annual household income reported that they lived in entrance-fee communities.

Monthly Fees

Nearly two-thirds of respondents in age-qualified communities said they paid monthly community fees, compared with 14% of residents of all-age communities. Among home owners who paid these assessments, the average fee was $219 in all-age communities and $295 per month in age-qualified communities. Among renters, those in

all-age communities said their monthly rent averaged $810 per month and those in age-qualified communities said they paid an average of $1,098 per month.

The greatest proportion of householders that paid these fees said the payments were for community maintenance, the right to occupy the residence, use of the amenities, and maintenance of roads. Close to equal proportions of home owners in all-age (45%) and age-qualified (43%) communities said their monthly fees paid for exterior home maintenance. A greater proportion of householders in age-qualified communities than those in all-age communities said their monthly fees paid for use of a golf course, interior home maintenance, trash removal, telephone connection, basic cable service, a security system, an emergency response system, Internet connections, dining services, transportation, housekeeping services, and social programs.

A Lifetime of Homeownership

Nearly half of 45+ middle-American heads of households have owned one (23%) or two (26%) homes. Nineteen percent have owned 3 homes, 15% have owned 5 or more homes, and 11% have owned 4 homes.

The number of homes owned during a lifetime increases slightly as the age of the head of the household increases (fig. 3.8), but the salient difference in number of homes owned is between the 45- to 54-year age-group and all other 45+ age-groups. Householders ages 45–54 have owned 2.3 homes, those 55–64 have owned 3.4 homes, and respondents 65–74 and 75+ said they have owned 3.1 homes. In other words, leading-edge boomers have owned as many homes in their lifetimes as respondents ages 75 and older.

As annual income increased, so did the number of homes owned. Heads of households with less than $70,000 in annual income reported owning an average of 2.6 homes during their lifetimes, compared with 4 or more homes for nearly 40% of home owners with annual incomes of $150,000 or more.

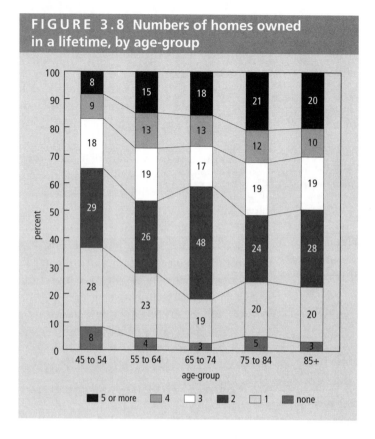

FIGURE 3.8 Numbers of homes owned in a lifetime, by age-group

likely to own 2 or more homes. Although the survey was not designed as a trend study, anecdotal evidence suggests that dual (or more) home owners may be a growth market. Thirty-five percent of respondents in the $150,000+ income group owned 2 or more homes, and 12% owned 3 or more homes. In contrast, less than 20% of those who earned less than $150,000 annually owned 2 or more homes, and less than 4% owned 3 or more homes.

Regionally, the greatest percentage of respondents that owned 2 or more homes was 18.6% in the Mid-Atlantic (ZIP Code Region 2) and 17.3% in the Southeast (ZIP Code Region 3).

The greatest proportion of heads of households with 2 homes said they purchased their

Slightly less than 40% of the home owners with annual incomes of $150,000 have owned 4 or more homes in their lifetimes, and 23% of respondents with incomes of less than $30,000 (again, these are likely people in the oldest age-groups) have owned 4 or more homes.

Home owners in the Southeast, the Mid-South and the Western Mountain regions (ZIP Code regions 3, 7, and 8) were significantly more likely to have owned five or more homes than the householders in other ZIP Code regions. Respondents in the Northeast, Pennsylvania, New York, and the Midwest (ZIP Code Regions 0, 1, 5, and 6) were the most likely to have owned 3 or fewer homes.

A majority (74%) of respondents owned one home at the time of the survey, but 14% owned 2, 3, 4, or more homes (fig. 3.9). Heads of households reporting $150,000 or more in annual household income are significantly more

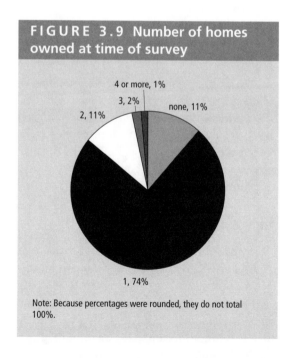

FIGURE 3.9 Number of homes owned at time of survey

Note: Because percentages were rounded, they do not total 100%.

second home as a vacation home (31%) or for investment purposes (30%). Almost 14% said they originally purchased their second home as their primary home, and 14% purchased their second home as a part-time residence. Heads of households who owned 3 or more homes were most likely to have purchased these additional homes for investment purposes. Several respondents with time-share properties counted those as their third (12%) and fourth homes (10%).

Let Buyers Lead the Way

With the number of homes owned in a lifetime apparently on the rise, U.S. housing consumers ages 45+ are likely to be much more discerning customers than might have been the case previously. These buyers have had ample opportunity to evaluate multiple homes for curb appeal, design flaws, beauty, and livability. Therefore, before beginning construction, builders should allow prospective home buyers the opportunity to examine possible floor plans and make changes to suit their lifestyles.

Moreover, recognizing the trend of homeownership extending well into people's 80s, many developers of age-qualified communities that provide a services package including services such as dining, housekeeping, transportation, and social programs, have begun to transition from rental or entrance-fee communities to offering condominium ownership.

PART II

WHERE THEY LIVE

4

Home, Location, and Satisfaction With Current Residence

Push factors move householders from their homes, and pull factors entice them to purchase a new home. Therefore, understanding where 45+ households live today, and in what types of homes, will help builders and developers determine how their communities can offer value-added designs, features, and amenities. This chapter examines, specifically, the types of homes survey respondents currently live in, how long they have lived in them and, most important, whether they like where they live. The data suggest that buyers 45–54 may be primed for a move—to a different home, neighborhood, or town or city.

The Location

The largest proportion (48%) of survey respondents said they lived in an outlying suburb, and about one-quarter said either they were in urban or in close-in suburban areas. For the purposes of the survey, urban was defined as within the central area of the town or city, and rural was defined as outside the town or city. Figure 4.1 shows the location of respondents' homes according to their age-group.

Among urban respondents, the percentage of renters was nearly twice that of home owners. As might be expected, 45+ middle-American householders with less than $30,000 in annual income were more likely to live in urban (28%) or sparsely populated rural areas. Thirty-nine percent of the householders with an annual income

of less than $30,000 lived in or near towns with populations of less than 10,000.

Meanwhile, the proportion of 45+ middle-American householders who live in areas with populations of less than 10,000 decreases with increasing income. Only 16% of the householders whose incomes were $150,000 or more lived in an area in which the closest town was that small; instead, almost a quarter said they lived near cities with populations of 500,000 or more.

Population concentrations differ throughout the country. Respondents in towns with smaller populations were in ZIP Code Region 0 (the Northeast), where 50% of the respondents said the nearest town had fewer than 25,000 people, and ZIP Code Region 5 (the Northern Plains), where 50% lived near a town of 10,000 or fewer residents. In contrast, householders in ZIP Code Region 9 (the Pacific Coast states) were the most likely (21%) to say they lived in an area with a population of 500,000 or more.

Communities: Conventional Neighborhood, Master-Planned, or Other

Survey respondents were provided with definitions for some common types of communities and asked to identify those that best described where they lived. The majority of respondents (56%) said they lived in conventional neighborhoods that were not part of a master-planned community. The rest were divided fairly evenly into two other types of communities:

- suburban master-planned communities that included homes and commercial establishments, offices or employers, and educational and community facilities conveniently located (23%); or
- other types of communities that were not part of a conventional neighborhood or a master-planned community (21%).

Only small proportions lived in age-qualified (5%), golf course (3.6%), or resort (1.4%) communities, although about 8% of householders who had lived in their homes 2 or fewer years were in golf course communities.

Of the small percentage of householders who lived in age-qualified communities, two-thirds were in active adult communities (67%). Other types of age-qualified communities and the percentage of respondents who lived in them are shown in figure 4.2.[2]

Gated Communities

Only a small minority (3%) of 45+ householders live in gated communities, although the percentage more than doubled (8%) for home owners who had been in their current residences for 2 or fewer years, for those with homes valued at $400,000 or more, and for those in ZIP Code Region 3. The percentage of the highest income respondents—those with $150,000 or more in

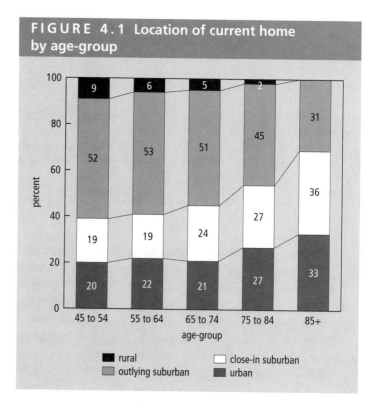

FIGURE 4.1 Location of current home by age-group

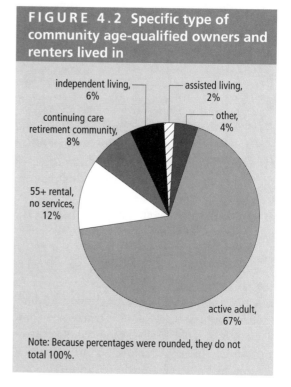

FIGURE 4.2 Specific type of community age-qualified owners and renters lived in

independent living, 6%

assisted living, 2%

continuing care retirement community, 8%

other, 4%

55+ rental, no services, 12%

active adult, 67%

Note: Because percentages were rounded, they do not total 100%.

---

[2]Respondents in assisted living communities are likely to be underrepresented in this survey because these individuals are frail, and some of them may not have telephones in their apartments.

annual income—who were living in gated communities was twice that of 45+ households as a group, or 6%. More householders 55+ years old (approximately 4%) live in gated communities than do 45- to 54-year-old householders (1%).

The factor that most influenced whether respondents ages 45+ were living in a gated community was whether they were in an age-qualified community. Nineteen percent of respondents in age-qualified communities were in gated communities.

Types of Homes

Although the single-family detached home continues to be the most common type of residence for 45+ households (87%), the proportion of 45+ middle-Americans moving to attached homes has steadily increased during the past 15 years. Of respondents who had been in their homes for 2 years or less, 12% were in attached housing

(fig. 4.3). Still, most respondents—even those in active adult communities—were in single-family homes.

The Mid-Atlantic had the highest percentage of households living in attached and multifamily residences (fig. 4.4).

Square Footage

The survey results confirm what other housing studies and census data have found—that home size has steadily increased since the 1950s. Forty percent of respondents ages 45+ whose homes were less than 2 years old at the time of the survey said their homes were 2,000–2,499 sq. ft., but only about a quarter of respondents who purchased their homes 2 to 10 years prior to the survey had homes that large.

Moreover, the proportion of respondents who were very satisfied with their homes increased with the home's square footage. Twenty-nine per-

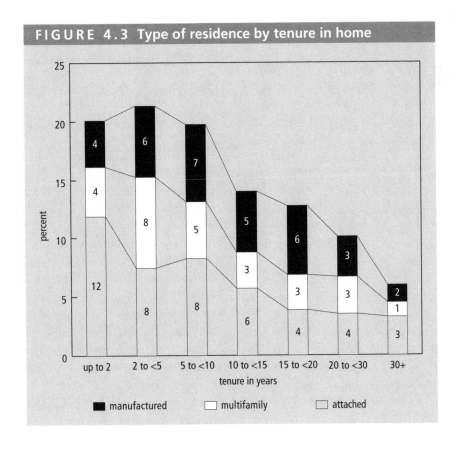

FIGURE 4.3 Type of residence by tenure in home

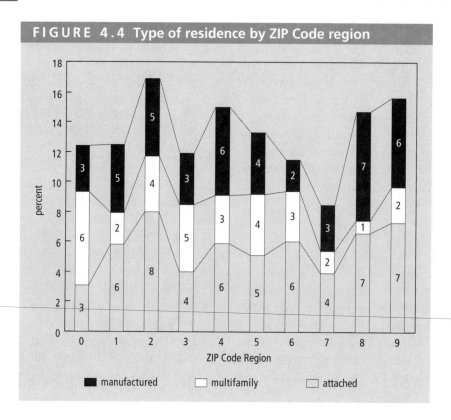

FIGURE 4.4 Type of residence by ZIP Code region

cent of home owners ages 45+ whose homes were less than 1,000 sq. ft. were very satisfied with their homes compared with 69% of respondents whose homes were 3,500–3,999 sq. ft. and 66% of those whose homes were 4,000 sq. ft. or larger.

Roughly equal percentages of home owners said their homes were 1,000–1,499 sq. ft. (22%), 1,500–1,999 sq. ft. (27%), 2,000–2,499 sq. ft. (21%), or 2,500 sq. ft. or larger (25%). Only 5% of middle-American home owners ages 45+ were in homes of less than 1,000 sq. ft. Reductions in home size were not apparent until respondents had reached age 85 or older.

As might be expected, home size increased with annual income, and home values were higher for larger homes.

Homes appear to be the largest in the Northeast (Zip Code Region 0) and the Mid-Atlantic (Zip Code Region 2), where 20% and 21%, respectively, are more than 3,000 sq. ft. In the Pacific Coast states (Zip Code Region 9), only 8% exceed 3,000 sq. ft. (fig. 4.5).

Number of Bedrooms

On average, survey respondents had 3 bedrooms, but the numbers ranged from none to 12. Only 0.3% had 7 or more bedrooms. The home size was proportionate to the number of bedrooms.

Half of middle-American home owners ages 45+ had 3 bedrooms in their residences, and 31% had 4 or more bedrooms. The rest had 2 (18%) or 1 bedroom (1%).

Among respondents who were 55+ years old, those in active adult communities were significantly more likely to live in 2-bedroom homes (62%) than those who lived in all-age communities (19%). Thirty percent of respondents who were 55 and older and living in active adult communities had 3-bedroom homes, compared with 51% of their counterparts who lived in all-age communities (fig. 4.6).

The proportion of 45+ middle-American home owners with 2 and 3 bedrooms decreases with increasing income, and the proportion with

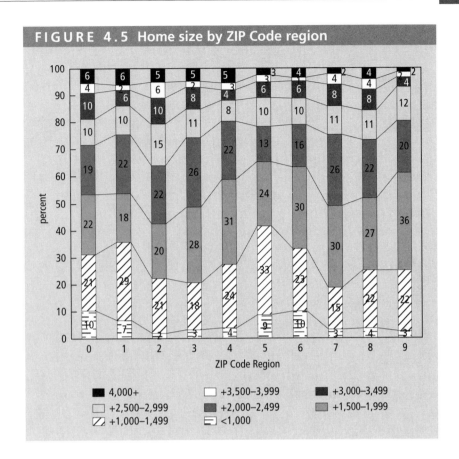

FIGURE 4.5 Home size by ZIP Code region

4- and 5-bedroom homes increases with increasing income. Approximately 50% of home owners who reported annual income of less than $30,000 to $99,999 had 3-bedroom residences. When households reached $100,000–$149,999 in annual income, 46% had 4 or more bedrooms. When income was $150,000 or more, 59% had 4 or more bedrooms.

Not surprisingly, higher home values corresponded with more bedrooms as well. Fifty-seven percent of the homes valued at less than $150,000 had 3 bedrooms, whereas 49% of the homes valued at $400,000+ had 4 or more bedrooms.

Number of Bathrooms

The greatest proportion of 45+ middle-American home owners reported having 2 full bathrooms in their residences (38%), but 18% had 2½ bath-

rooms, 16% had 1, and 13% had 1½ bathrooms. Fifteen percent had 4 or more bathrooms.

The greatest proportion of each age-group had 2 bathrooms. The age-groups between 45 and 64 were slightly more likely than householders ages 65+ to have more than 2 bathrooms.

Three-fourths of home owners in active adult communities had 2 bathrooms in their homes. In contrast, 36% of home owners ages 55+ who lived in all-age communities had 2 bathrooms (fig. 4.7). The 55+ home owners in all-age communities were more likely to have 1 or 1½ bathrooms (31%) than those who lived in active adult communities (14%). Thirty-two percent of home owners in all-age communities had more than 2 bathrooms, compared with 11% of home owners who lived in age-qualified communities.

The majority of home owners ages 55+ in active adult communities with 2-bedroom

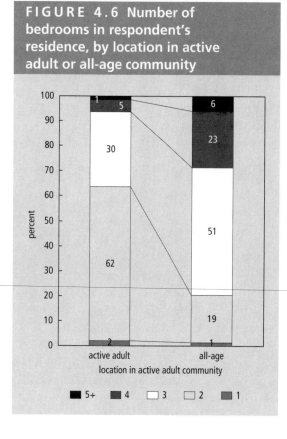

FIGURE 4.6 Number of bedrooms in respondent's residence, by location in active adult or all-age community

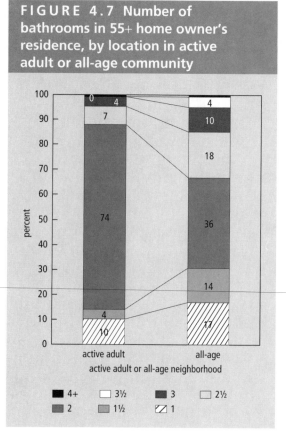

FIGURE 4.7 Number of bathrooms in 55+ home owner's residence, by location in active adult or all-age community

homes had 2 bathrooms (77%), and 82% with 3-bedroom homes also had 2 bathrooms. Although a majority of the 55+ home owners in all-age communities with 2 bedrooms had 2 bathrooms (44%), they were more likely than their counterparts in active adult communities to have 1 bathroom (32%) or 1½ bathrooms (19%). Similarly, a majority of 55+ home owners in 3-bedroom homes in all-age communities had 2 bathrooms (44%), but 17% had 1 bathroom, and 15% had 1½ bathrooms.

Forty-eight percent of 45+ households with less than $30,000 in annual income had 1½ or fewer bathrooms. In contrast, 4% of the home owners with $150,000 or more in annual income had 1½ or fewer bathrooms.

As the home values increased, the proportion of homes with more than 2 bathrooms did as well. Eleven percent of homes valued at less than $150,000, 45% valued between $200,000 and $299,999, and 62% of the homes valued at $400,000+ had more than 2 bathrooms.

Where incomes and home values or home sizes were the largest, so were the number of bedrooms and bathrooms. Respondents in the Mid-Atlantic, Southeast, Mid-South, Western Mountain, and Pacific Coast states (ZIP Code Regions 2, 3, 7, 8, and 9) were more likely to have 2 or more bathrooms than were residents in the Northeast, New York and Pennsylvania, Ohio Valley, Northern Plains, or the Midwest (ZIP Code Regions 0, 1, 4, 5, and 6).

Number of Stories

Not including basement level, 63% of middle-American home owners ages 45+ who lived in single-family residences lived in single-story homes; 29% lived in 2-story homes; 5% lived in split-levels; and 3% lived in 3-story residences.

Fifty-three percent of respondents ages 45–54 lived in a single-story home compared with 73% of householders 75+ years old. The proportion of households in split-level and 3-story homes was fairly constant across age-groups, whereas the proportion that lived in 2-story homes declined as age increased. Thirty-eight percent of 45- to 54-year-olds, compared with 19% of respondents who were 75+, lived in 2-story homes.

More 55+ home owners who lived in active adult communities were in single-story homes, compared with home owners in all-age communities—88% versus 65%. Only 8% of the 55+ home owners in active adult communities lived in 2-story homes, compared with 27% of this same age-group who lived in all-age communities.

The largest percentages (60% and 65%, respectively) of 45+ middle-American householders who lived in multistory homes were in the Northeast and New York/Pennsylvania. The smallest proportions were in the Mid-South and Southeast. Where high proportions of people live in multistory homes, population is denser and thus, land costs more. In areas where there are lower population densities, land prices are lower and more people live in single-story homes.

Stairs to Bedrooms

One of the primary designs potential buyers who are ages 55+ ask for is a one-story home, or at least a home that has the owner's suite on the ground floor. Survey responses show that builders have responded to this demand. During the past 15 years, the proportion of home owners ages 45+ that own homes requiring them to climb stairs to reach their bedrooms has steadily declined. Among those in active adult communities, only 5% said they had to climb stairs to reach their bedrooms, compared with 22% of home owners 55 and older who were in all-age communities.

Among all survey respondents, 25% of middle-American home owners ages 45+ had their bed-

rooms on a floor that required climbing stairs. The proportion that had to climb stairs to reach their bedrooms decreased as age increased (fig. 4.8). Thirty-four percent of home owners ages 45–54 and 29% of home owners ages 55–64 said they climbed stairs to reach their bedrooms, compared with roughly 15% of householders who were 75+ years old.

> *Single-level living is a must. The houses need to have the ability to add loft space to create vertical separation within the unit. With the advent of the baby boomers, the units need to contain volume and be open.*
>
> **—WILLIAM P. GERALD, JR.**
> Vice President
> Classic Community Companies
> Bethesda, MD

Overall, it appears that more multistory homes are being built today with the owner's suite on the ground floor. Thirty-seven percent of the 45+ householders who reported the age of their homes at 50 years or older said they had to climb stairs to reach their bedrooms. This category of homes may include some built in the 1700s. When isolating homes built in the past 50 years, though, the highest proportion that required climbing stairs to reach bedrooms appears to be in homes built in the past 16–25 years.

Forty-six percent of respondents in the Northeast (ZIP Code Region 0), 55% of respondents in New York and Pennsylvania (ZIP Code Region 1), and 37% of respondents in the Mid-Atlantic (ZIP Code Region 2) had to climb stairs to reach their bedrooms (fig. 4.9).

Basements

About half (53%) of 45+ middle-American home owners said their homes had basements. Only 22% of 55+ householders in active adult communities had basements, although 53% of the 55+ group in all-age communities did.

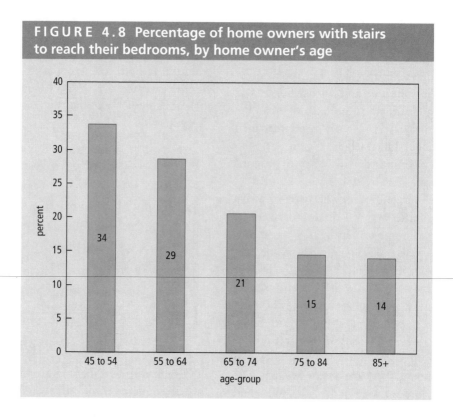

FIGURE 4.8 Percentage of home owners with stairs to reach their bedrooms, by home owner's age

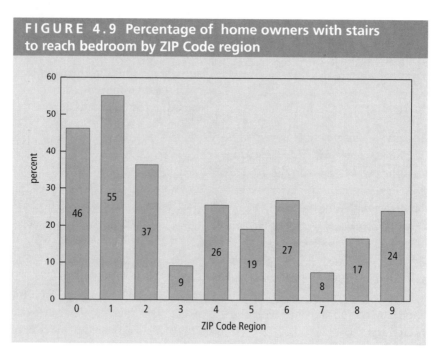

FIGURE 4.9 Percentage of home owners with stairs to reach bedroom by ZIP Code region

Respondents living in homes that were 50 or more years old were also more likely to have basements. For example, 73% of home owners who were ages 45+ and in homes 50 or more years old had basements, compared with 47% of those in homes 16–20 years old and 32% of those in homes that were less than 2 years old when the survey was conducted.

Figure 4.10 clearly shows also that preference for basements (or not) varies by region.

Satisfaction with Homes, Neighborhoods, and Towns

Home owners have a greater sense of satisfaction with the aspects of their living environments over which they have the greatest control. Half of the 45+ middle-American home owners were very satisfied with their homes, 45% were very satisfied with their neighborhoods, and 37% were satisfied with the towns or communities in which they lived.

The proportion of respondents that was very satisfied with their homes, neighborhoods, or towns increased steadily and significantly with age. For example, only 36% of 45- to 54-year-olds, compared with 63% of respondents who were 85+, were very satisfied with their homes. The same pattern emerged when looking at respondents' satisfaction ratings for their neighborhoods and towns (fig. 4.11).

> *The competition you face is not other communities in your marketplace. Your true competition is the homes they are in now.*
>
> —**ROBERT H. KAREN**
> Managing Member
> Symphony Development Group, LLC
> Owings Mills, Maryland

Middle-American householders with high incomes had a greater sense of satisfaction with their homes and neighborhoods than middle-American householders with low incomes, but the differences between these two groups were not as pronounced for their satisfaction ratings of their towns and communities.

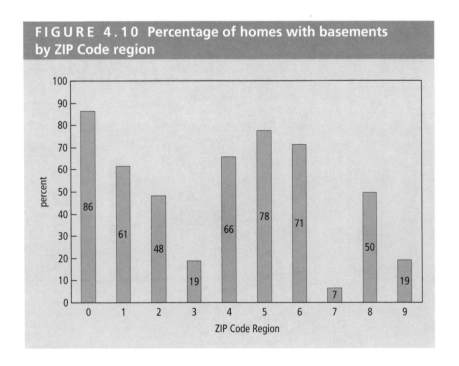

FIGURE 4.10 Percentage of homes with basements by ZIP Code region

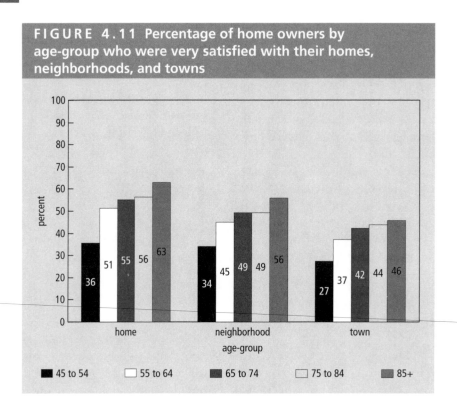

FIGURE 4.11 Percentage of home owners by age-group who were very satisfied with their homes, neighborhoods, and towns

The proportion of 45+ householders that was very satisfied with their homes or neighborhoods also increased steadily with home value. Forty percent of householders with homes valued at less than $150,000, compared with 61% of householders who estimated their home's value at $400,000 or more, were very satisfied with their homes. Roughly 40% of respondents with

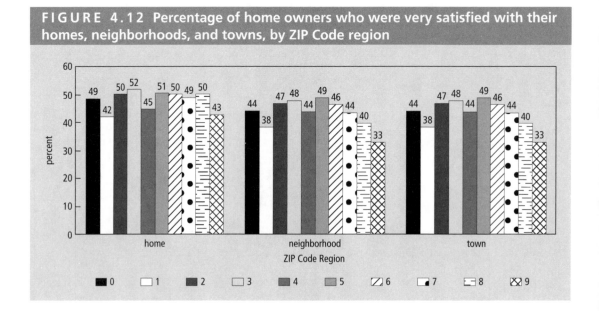

FIGURE 4.12 Percentage of home owners who were very satisfied with their homes, neighborhoods, and towns, by ZIP Code region

home values of $150,000–$400,000+ said they were very satisfied with their towns, compared with 30% of respondents whose home values were less than $150,000.

Households appear to be the happiest with their homes, neighborhoods, and towns in the Southeast (ZIP Code Region 3), Northern Plains (Region 5), and the Mid-Atlantic (Region 2). New York and Pennsylvania (ZIP Code Region 1), the Ohio Valley (Region 4), and the Pacific Coast states (Region 9) have the smallest proportion of households who are very satisfied with their homes, neighborhoods, and towns (fig. 4.12).

Pulling in 45+ Buyers

Push factors that encourage 45+ buyers to seek out a new residence include size of home (too large or too small), stairs, or dissatisfaction with a neighborhood or community. Pull factors include a community in a desirable location and a new home that is all on one level.

5

45+ Renter Households

Less than 10% of 45+ middle-American householders are renters, and most of these households are headed by a person who is either 45–54 or 75+ years old. Members of the latter group have the longest tenure in their rental properties, compared with other age-groups of 45+ renters. Those in the former group are still in a stage of life in which they are likely to purchase homes, but 75+ renters have reached a stage in which they may feel the need to divest themselves of the responsibilities that come with home-ownership. Members of the latter group have the longest tenure in their rental properties.

Geographically, renters are more likely to be in urban and close-in suburban areas where land is scarcer and homes, consequently, are more expensive. Forty-three percent of 45+ middle-American renters live in urban areas, 27% live in close-in suburban areas, 26% live in outlying suburban areas, and a mere 4% live in rural areas.

Location

Similar to the pattern seen among home owners, renters ages 45–54 and those who are 85+ both are more likely to live in urban areas than their counterparts in other 45+ age-groups. The proportion in urban and close-in suburban areas increases among householders ages 75 and older. When people require outside assistance with day-to-day living, they move toward more services and people they know.

More than half (57%) of renters with the lowest incomes (less than $30,000 annually) live in urban areas. This group includes a large proportion of 85+ householders who are likely to be single women. Renters with $100,000+ in annual income are more likely to be found in rural areas than are rental households with lower incomes.

Almost half of the 45+ renter householders in the Northeast (49%) live in an urban area, compared with 34% of the renters in the West. In many areas in the Northeast land costs are high, and many areas with a high population density have little if any vacant land. This situation impacts the number of affordable homes, or even whether any homes can be built in some locations. Where land is scarce and home prices are high, more 45+ households are renters.

Tenure in Home

On average, renters who are 45+ live in their residences for 7.2 years. Older renters have lived in their current residences longer, and younger renters have been in their homes for a shorter period of time. Renters ages 55+ have lived in their homes for 7.9 years on average, and no difference exists in the tenure of residency between those who live in age-qualified communities and those who live in all-age communities.

Type of Community

Significantly more renter householders ages 45+ (8%) than owner households (4%) live in age-qualified housing. Still, with the overwhelming preference for homeownership among 45+ households, the percentage of renters in age-qualified communities is small. Approximately the same proportion of renters as home owners lived in gated and golf course communities, but a higher proportion of renters (8%) said they lived in resort communities when compared with owners (5%).

Types of Age-Qualified Communities

Renters identified the types of age-qualified communities they lived in. The largest proportion lived in age-qualified, multifamily apartment buildings (42%) that typically had fitness centers and common areas for residents, but these buildings did not include a central kitchen or dining facility. Twenty-five percent lived in active adult communities; 15% lived in continuing care retirement communities; and 13% lived in independent living communities (fig. 5.1).

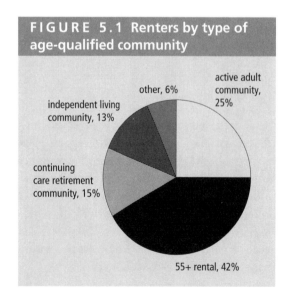

FIGURE 5.1 Renters by type of age-qualified community

other, 6%

independent living community, 13%

active adult community, 25%

continuing care retirement community, 15%

55+ rental, 42%

Types of Residences Rented

Forty-nine percent of renters ages 45+ lived in multifamily housing, 37% rented detached homes, 11% rented attached homes, and 4% rented manufactured homes. As the age of the renters increased, so did their likelihood of living in multifamily buildings or attached homes. Forty-five percent of renters ages 45–54 lived in detached single-family homes, compared with 22% of 75+ renters. The percentage of renters living in attached homes increased from 5% among 45- to 54-year-olds to 16% of the 75+ age-group. The percentage of renters in multifamily housing increased from 48% to 58%, respectively, for those age-groups (fig. 5.2).

Number of Stories in Multifamily Buildings

Only 14% of renters ages 45+ lived in a high-rise building with 6 or more stories. About half (49%) of all 45+ middle-American renters who lived in multifamily housing lived in 2-story buildings. The rest lived either in 3-story buildings (26%) or buildings with 4 or more stories (25%).

Interestingly, 74% of renters ages 45+ who lived in buildings with 2 or more stories did not have elevators. In all, 93% of those who lived in 2-story buildings, 79% in 3-story buildings, 40% in 4-story buildings, and 67% in 5-story buildings (8 of 12 total respondents) did not have elevators. All respondents with 6 or more stories in their buildings reported that their buildings had elevators. The proportion of renters that lived in buildings with 2 or more stories that had elevators increased with the renter's age. Still, only 43% of respondents ages 75+ said they had elevators in their multistory buildings.

One-fourth of the renter householders with less than $30,000 in annual household income, compared with two-thirds of those who had $150,000+ in annual income, said their multistory buildings had elevators.

Two of the most prevalent complaints among mature adults are difficulties with their knees and

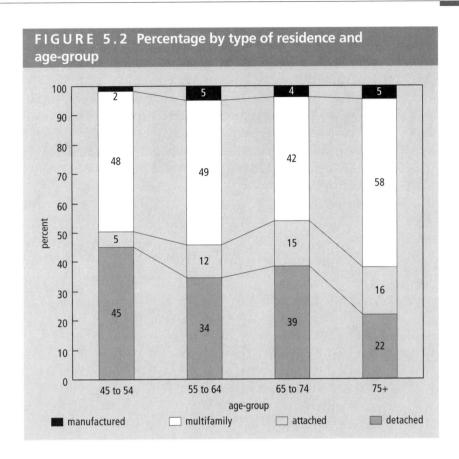

FIGURE 5.2 Percentage by type of residence and age-group

manufactured   multifamily   attached   detached

loss of strength and stamina. Many of the householders living in buildings without elevators will be forced to move eventually.

Sizes of Residences

Among middle-American renters ages 45+ who lived in multifamily buildings, 33% were in residences of 800–999 sq. ft., 25% were in residences of 1,200 sq. ft. or more, 22% were in homes of 1,000–1,199 sq. ft., and 20% lived in residences that were less than 800 sq. ft.

Although the square footage of rentals differed across age-groups, those differences did not follow a discernible pattern. Higher proportions of both 45- to 54-year-olds and respondents who were 75+ lived in multifamily residences of 800–999 sq. ft. A greater proportion (25%) of householders ages 75+ than those ages 45–54

(17%) lived in residences of 1,200 sq. ft. or more. As was the case with home owners, a larger proportion of 55- to 64-year-old renters than other age-groups had the largest residences. Seventy percent of 55- to 64-year-old renters lived in residences that were 1,000 sq. ft. or more. Forty-three percent or less of the remaining age-groups rented residences of 1,000 sq. ft. or larger (fig. 5.3).

Renter householders ages 55+ in age-qualified communities were significantly more likely to live in residences of 800–999 sq. ft. (69%) than were 55+ householders in all-age communities (23%). The largest proportion of renters ages 55+ in all-age communities (34%) lived in residences that were 1,200 sq. ft. or larger.

Of the multifamily renters, households with incomes of $100,000+ were the most likely to have rental residences of more than 1,200 sq. ft.,

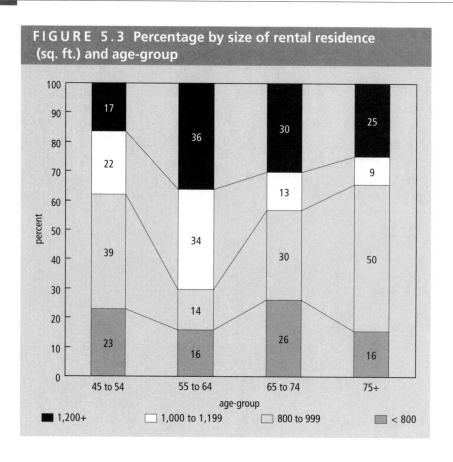

FIGURE 5.3 Percentage by size of rental residence (sq. ft.) and age-group

but, with the exception of the lowest income groups, the wealthiest group also was the most likely to be living in residences of less than 800 sq. ft. (fig. 5.4).

No statistically significant differences were apparent in incomes of renters in age-qualified communities versus those in all-age communities. Approximately 62% of renters ages 55+ said their incomes were less than $30,000. A slightly larger proportion of householders in age-qualified communities than in all-age communities had incomes above $100,000, but the difference was not statistically significant.

Renters in the Northeast were the most likely to live in multifamily buildings in apartments of less than 800 sq. ft. (38%). Renters in the South were the most likely to live in residences of 1,200 sq. ft. or more (40%). Renters in the West were the most likely to live in apartments of 800–999 sq. ft. (61%) (fig. 5.5).

Bedrooms and Bathrooms

More than half (61%) of renters ages 45+ in multifamily buildings had 2-bedroom residences; 32% had 1 bedroom; and 7% had 3 or more bedrooms. These findings held for renters in age-qualified and all-age communities alike. The number of bedrooms differed according to the respondent's age, however. For example, although 44% of 55- to 64-year-old renters were in 1-bedroom homes, in the 3 other age-groups the proportion was 30% or less.

The vast majority of 45+ middle-American renters in multifamily buildings said their apartments had 1 (67%) or 2 (26%) bathrooms.

Length of Residence

Among 45+ renters the largest percentage (40%) had lived in their residences for less than 2 years, although tenure varied by type of residence.

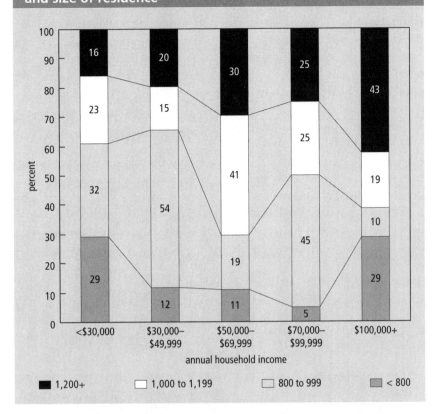

FIGURE 5.4 Percentage by annual household income and size of residence

(legend) ■ 1,200+ □ 1,000 to 1,199 □ 800 to 999 ■ < 800

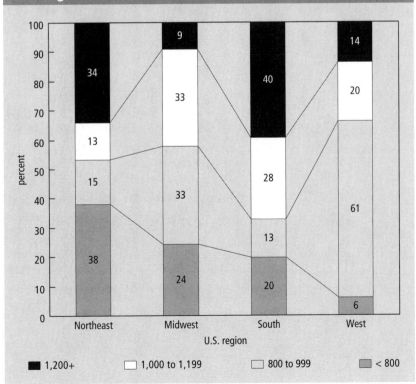

FIGURE 5.5 Percentage by residence size and U.S. region

(legend) ■ 1,200+ □ 1,000 to 1,199 □ 800 to 999 ■ < 800

Renters in manufactured housing had the shortest tenure; 65% had lived in their homes for less than 2 years and none had lived in them more than 15 years (fig. 5.6).

Renters in multifamily housing appear to have the longest tenure in rental housing. Although 30% had lived in their apartments for less than 2 years, 27% had lived in them for more than 10 years.

Survey results showed no statistically significant differences in the length of time renters had lived in their residences (whether single-family, multifamily, or manufactured) by whether or not the residence was in an age-qualified or an all-age community.

Compared with other regions, renters ages 45+ in the Northeast were likely to have lived in their residences longer. Forty-three percent had lived in their rental residences 10 years or longer. Renters

in the South and the West had the shortest tenures in their residences: 43% of those in the South and 41% of those in the West had lived in their homes for less than 2 years (fig. 5.7).

Satisfaction with Homes and Neighborhoods

A significantly smaller percentage of renters, compared with home owners, was satisfied with their residences and neighborhoods. However, as with home owners, the percentage of renters that was satisfied with their homes, neighborhoods, and towns was lower for respondents in the 45–54 age-group than for the other groups (fig. 5.8).

Higher incomes appear to increase satisfaction. Thirty-nine percent of renters ages 45+ with $100,000+ in annual household income said they

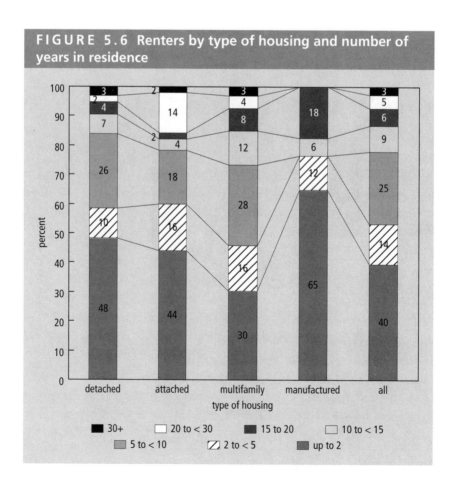

FIGURE 5.6 Renters by type of housing and number of years in residence

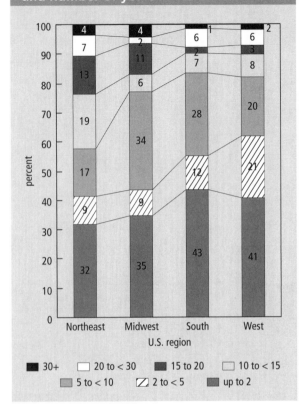

FIGURE 5.7 Percentage by region and number of years in residence

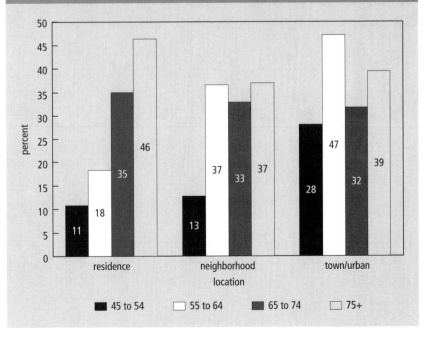

FIGURE 5.8 Percentage of renters, by age-group, who were very satisfied with their homes, neighborhoods, and towns

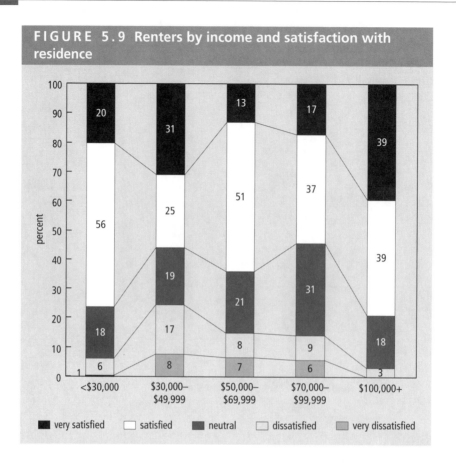

FIGURE 5.9 Renters by income and satisfaction with residence

were very satisfied with their rental residences (fig. 5.9). The proportion declined with income, but the relationship of income to satisfaction was not linear: respondents with incomes of $30,000–$49,999 were the least satisfied.

Planning for 45+ Renters and Buyers

The highest proportion of renters among middle-American households ages 45+ are in the 45–54 and 75+ age-groups. Although 45- to 54-year-olds are still likely to be potential home buyers, 75+ renters have entered a life stage in which they may want to divest themselves of homeownership. Among renters surveyed, those who were 75+ had the longest residency in their current rental properties.

As might be expected, renters were more likely to be in urban and close-in suburban areas where land is more expensive and, therefore, homes are as well. The greatest proportion of renters ages 45+ had 2 bedrooms, but nearly one-third had 1 bedroom.

6

Movers: Those Planning to Move and Those Who Moved Recently

The survey results and my own personal experience demonstrate that when planning current and future projects, builders and developers should not underestimate Americans' attachment to their current homes.

About 10 years ago I partnered in developing an independent living and assisted living community where I live in Oxford, Miss. I fantasized that my parents might one day relocate to this community. Therefore, I kept in mind what I thought my mother would want. But a decade went by, my father passed away, and my mother had a new home built just for her in Illinois.

After my mother wintered down South near me in a second home that I own, we looked forward to her returning to the home permanently. But instead she had her own winter home built in Mississippi. Now, she'll be able to spend time near us, but in her own house.

In fact, the vast majority of middle-Americans ages 45+ have a strong desire to live in their own homes. Still, a comparison of the survey responses to questions about whether or when respondents might move with results of other studies indicates that many underestimate their likelihood of moving. This chapter examines attitudes about moving and the strength of attachment respondents have to where they live, according to various aspects of their lives and the elements that influence their willingness and ability to move. It also explores differences between movers and stayers (those who never plan to move) and discusses the reasons

people gave for planning to move or making a move.

Moving

Multiple studies including the U.S. Census indicate that approximately 4% of all households move each year. The survey conducted for this book found that 8% of householders ages 45+ moved in the 2 years prior to the survey and 10% planned to move within 5 years. Nearly equal percentages of the remaining respondents said they did not intend to move from their current homes (42%) or that they would move in the future but were unsure of when they would do so (40%).

Almost half (47%) of 45+ middle-American renters said they were likely or very likely to move in the future. In contrast, 27% of the householders who owned their homes said they were likely to move in the future.

Attitudes That Influence the Decision to Move or Stay

Given that homeownership is the American dream, it is no surprise that survey respondents had an incredibly strong attachment to "my own house." Eighty-two percent of them agreed or strongly agreed with the statement, "I would prefer to live in my own house even if it meant that I would spend most of my time alone." Only 6%

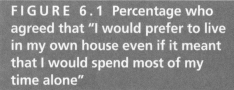

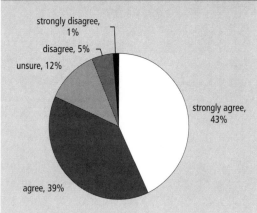

FIGURE 6.1 Percentage who agreed that "I would prefer to live in my own house even if it meant that I would spend most of my time alone"

of survey respondents disagreed with the statement (fig. 6.1).

Women were slightly more inclined to want their own homes than men. Forty-five percent of them, compared with 40% of the men, strongly agreed with the statement. Among respondents who were widowed, divorced, or never married, 55% of the women and 44% of the men strongly agreed that they would prefer to live in their own houses even if doing so meant spending most of their time alone.

The Decision to Move

An equal percentage of 45+ middle-American householders said they were very unlikely to move from their current primary home in the future (29%) as said that they were likely or very likely to move (28%). Twenty-two percent and 21%, respectively, said that they were unsure about whether they would move or that a move was unlikely. In other words, about half of householders ages 45+ contemplate moving from their current homes at some time. Even if those moves are staggered over decades, that's a lot of moving vans. In

addition, keep in mind that these statistics don't include those who plan to purchase a second (or third) home as a part-time, vacation, or temporary residence.

The likelihood of moving decreases with increasing age. Thirty-nine percent of the 45–54 age-group, compared with 31% of 55- to 64-year-olds, 22% of 65- to 74-year-olds, 9% of 75- to 84-year-olds, and 10% of the 85+ age-group said that they were likely or very likely to move in the future (fig. 6.2). One of the reasons that fewer older respondents said they will move may be because they already have moved.

As income increased, so did the proportion of respondents who said that they were very likely to move (fig. 6.3). In addition to perhaps having insufficient funds to make a move, the lower-income groups also may include a larger proportion than other income groups of 85+ households that already have moved.

Slightly more than one-third (34%) of householders ages 45+ with home values of $400,000+ are likely to move in the future, and householders with home values of less than $150,000 are the least likely to move in the future (23%).

The longer respondents lived in their homes, the lower the probability that they planned to move. Granted, people who lived in their homes longer also were likely to be older. Comparing the likelihood of moving within an age-group, though, reveals that people who have moved in recent years were more likely to move than those who had not moved in recent years. Thirty-nine percent of survey respondents who had lived in their homes for less than 2 years said that they were likely to move in the future, compared with 18% of those who had lived in their homes for more than 30 years. Figure 6.4 shows the likelihood of moving by ZIP Code region.

What Drives People to Move

Dissatisfaction with their current residences drives the majority of respondents to move. Sixty-one percent of those who were very dissatisfied and

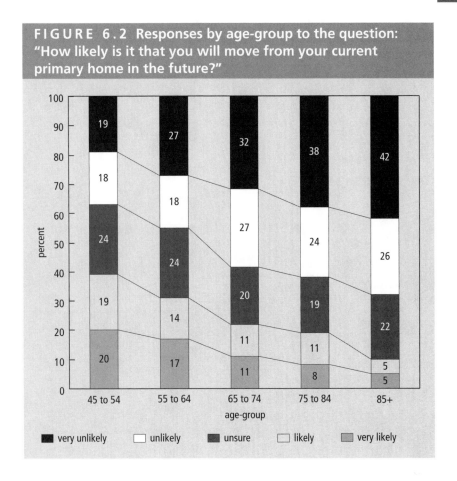

FIGURE 6.2 Responses by age-group to the question: "How likely is it that you will move from your current primary home in the future?"

50% who said they were dissatisfied with their current residences said they were very likely to move in the future. In contrast, 8% who were very satisfied said they were likely to move in the future. The same effect occurs among 45+ middle-American householders who are dissatisfied with their neighborhoods and towns. Householders who were dissatisfied with their neighborhoods or towns were significantly more likely to plan to move than those who were satisfied.

Comparing Recent Movers with Those Planning to Move

Either people who have moved have forgotten how long they planned to move before they actually did so, or those planning to move have no idea how quickly moves occur once the ball is set in motion. Thirty-six percent of the respondents who moved within the past 2 years said they planned or anticipated their move for fewer than 6 months prior to moving. In contrast, only 9% of householders who said they were planning to move believed they would move in fewer than 6 months. Only 12% of those who moved said they took three years to move after they first anticipated moving, whereas 54% of those planning to move said that it would be 3 years before they would move.

Reasons for Moving

The sections that follow compare the reasons for moving between two groups: those who moved in the past two years (identified in the accompanying figures as "moved") and those

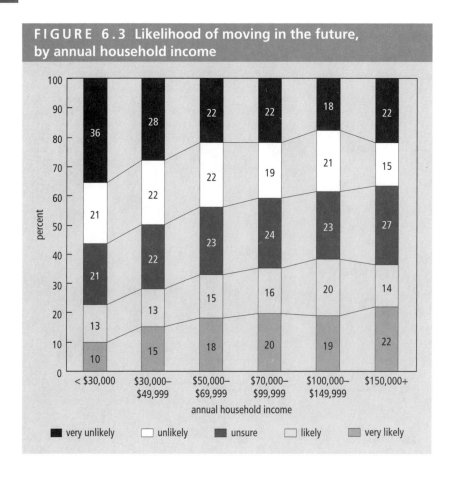

FIGURE 6.3 Likelihood of moving in the future, by annual household income

who are planning to move (identified in the accompanying figures as "planning"). Survey respondents identified their reason for moving by choosing from a series of statements about why they moved or planned to move. Their responses covered

- a primary reason for moving
- a partial reason for moving
- no reason for moving

Size and Design of Home

At least 61% of all home owners surveyed, regardless of their current home's value, said their desire for a low-maintenance home was part of their reason for moving. More householders ages 45+ (47%) who had moved or who were plan-

ning to move said having an accessible and easier-to-use home was motivating them than said they were looking for either a larger or a smaller home.

Among the group of 45+ middle-American householders who were planning to move or who had moved in the 2 years prior to the survey, a slightly greater proportion of the former group planning to move said they were looking for a smaller home (34%). Twenty-nine percent were looking for a larger home.

For some respondents, the desire for something new or for a change was their primary reason for moving. Thirty-six percent of those planning to move and 29% of those who had moved recently said one of the reasons for moving was to have a newly built home. The proportion of respondents who said that having a new

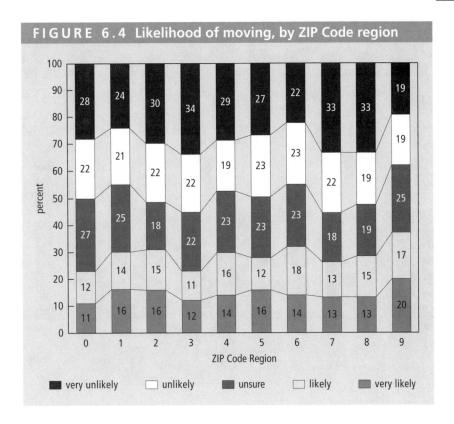

FIGURE 6.4 Likelihood of moving, by ZIP Code region

home was part of their reason for moving declined with increasing age: 18% of the 75+ age-group compared with 40% of the 45–54 age-group said the desire for a new home was part of the reason they were moving.

The householder's age also was one factor in size and design preferences. As age increased, a smaller proportion wanted a larger home. Only 9% of householders ages 75+ said the desire for a larger home was part of the reason they were moving (fig. 6.5). To the contrary, as age increased, a greater proportion of respondents said that part of the reason they were moving was to have a smaller home. More than a quarter of the 45–54 age-group also wanted a smaller home.

An accessible, easy-to-use home was part of the reason for moving for at least 40% of those 45–54, 51% of the 55–64 age-group, and almost 60% of householders ages 65+. Similarly, more than half of respondents ages 45–54 and more than 68% of 55+ householders said having a low-maintenance home was part of the reason they were moving.

Size of Lot

Apparently, householders planning a move are more likely to accept a home on a smaller lot, or to have no lot, when compared with householders who had moved in the 2 years prior to the survey. The largest proportion of 45+ householders planning to move (27%) or who had moved within the 2 years prior to the survey (26%) said the desire for a larger lot was part of the reason for their move (fig. 6.6). However, one-fourth of those planning to move and 15% of those who had moved said the desire for a smaller lot was part of the reason for their move. Thirteen percent of those planning a move and 8% of those who had moved said the desire for no lot was part of their reason for moving.

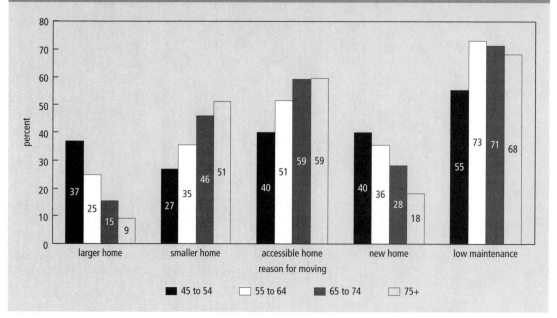

FIGURE 6.5 Percentage, by age-group, that said wanting a larger, smaller, accessible, new, or low-maintenance home was a primary or secondary reason for moving

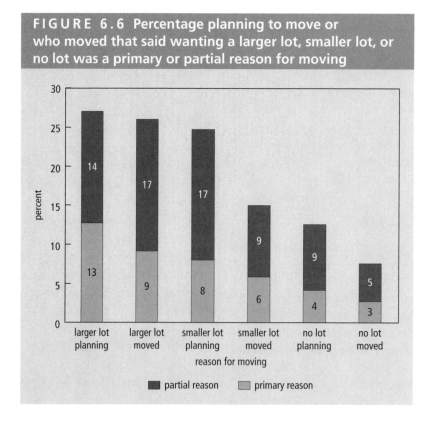

FIGURE 6.6 Percentage planning to move or who moved that said wanting a larger lot, smaller lot, or no lot was a primary or partial reason for moving

The proportion of respondents that wanted a larger lot decreased as their ages increased. The opposite was true for a smaller lot or no lot at all. Seventeen percent of respondents ages 45–54 wanted a smaller lot, compared with 28% or more of 55+ householders. Seven percent of the 45–54 age-group, 16% to 18% of the 55–74 age-group, and 30% of the 75+ age-group said wanting a home with no lot was at least part of the reason for their move.

Neighborhood and Community Attributes

Householders who had moved recently indicated that finding a better neighborhood was at least part of the reason for their move (37%). This was a greater proportion of householders than among those planning a move (27%). About equal proportions of those planning to move and those who had moved recently said part of the reason for their move was to live where they would have a better sense of security (fig. 6.7). Twenty-three percent planning to move and 15% who moved recently said amenities were part of the reason for their move.

More 45+ middle-American householders said they planned to move or had moved to be closer to family than said they did so to be closer to shopping or to their workplaces. Thirty-one percent who were planning to move wanted to be closer to family, 20% wanted to be closer to shopping, and 10% wanted to be closer to work.

Living in a better neighborhood was less important for householders ages 55+ compared with those ages 45–54 (fig. 6.8). Better security and being closer to family were part of the reason for moving for at least 23% of all age-groups, but a greater proportion of householders ages 75+ compared with younger householders said these were

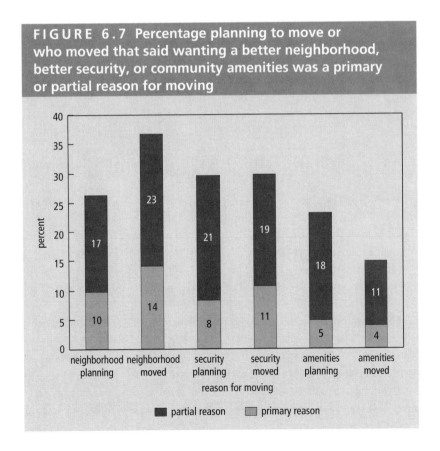

FIGURE 6.7 Percentage planning to move or who moved that said wanting a better neighborhood, better security, or community amenities was a primary or partial reason for moving

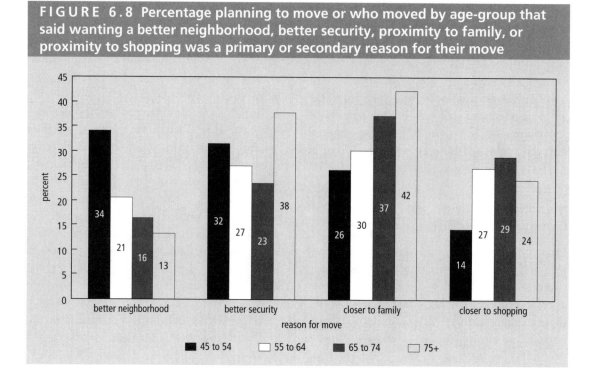

FIGURE 6.8 Percentage planning to move or who moved by age-group that said wanting a better neighborhood, better security, proximity to family, or proximity to shopping was a primary or secondary reason for their move

the primary or secondary reasons for their moves. At least 24% of householders ages 55+ compared with 14% of the 45–54 age-group said being closer to shopping was part of their reason for moving.

Stage of Life

A smaller proportion of householders who had moved compared with those who were planning to move stated that at least part of the reason for their move was a change in their or their spouse's health, change in the number of people living in their household, or their retirement (fig. 6.9). Almost half of the householders who were planning to move said that retirement was at least part of the reason they were moving; 30% said a change in health status and 33% said a change in household size influenced their decisions. Almost equal proportions of householders who were planning to move (10%) and who had moved (12%) said a change in marital status was part of

the impetus for their moves. Essentially equal proportions of those planning to move (13%) and those who had moved (14%) said part of their reason for moving was job related.

A change in health is one of the primary reasons prompting people 75+ years old to plan a move. Health also prompted moving plans for almost one-fifth of the 45–54 age-group, one-third of those 55–64, and 44% of those 55–64. Change in the number of people living in the house was more likely to be cited as a reason to move among households headed by people ages 45–64 than among those ages 65+. Sixty-four percent of survey respondents ages 55–64 who were planning to move said that retirement was at least part of the reason for their plans to move. Interestingly, 38% of householders in the 75+ age-group also said retirement was influencing their plans for a move. A move related to a job was part of the rationale for 10%–18% of respondents ages 45–64 but only 3% of those in the 75+ age-group.

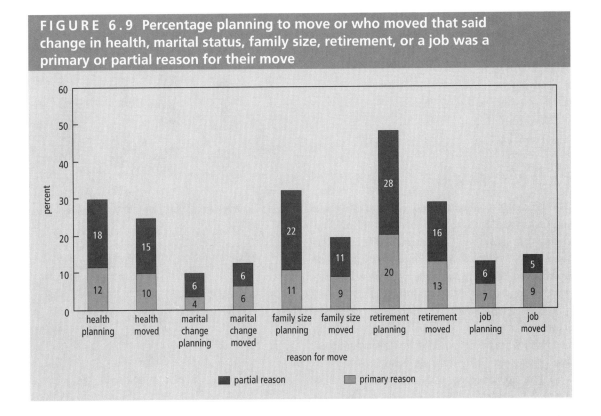

FIGURE 6.9 Percentage planning to move or who moved that said change in health, marital status, family size, retirement, or a job was a primary or partial reason for their move

Financial Considerations

Among householders planning to move, 31% said they wanted to lower expenses; 23% said they wanted to own their residences; 18% wanted to cash in their equity; 16% said part of the reason they were moving was for investment goals related to homeownership; and 15% said getting away from Home Owner Association (HOA) dues was part of their motivation to move. The greatest difference between those planning to move and those who had moved in the two years prior to the survey had to do with reducing expenses. Thirty-one percent who were planning to move, compared with 17% of those who had moved, said the reason for moving was to reduce expenses.

Financial reasons for moving differed by age-group. A significantly greater proportion of householders ages 45–64 (18%) than householders ages 65+ (less than 8%) said part of their

motivation to move was related to an investment opportunity. Similarly, more than one-fourth in the 45–64 age-group wanted to own their homes, compared with 14% or fewer of those ages 65+. A greater proportion of 75+ householders (34%) said part of their motivation to move was to cash in equity, compared with the proportion of householders in the 45–64 age-group (less than 24%).

Twenty-five percent of householders planning to move who estimated the value of their homes at less than $200,000 and 37% or more of the householders whose home values were $200,000+ said reducing expenses was part of their motivation to move. For homes valued above $200,000, the desire to cash in equity increased as the value of the homes increased.

When householders in the 45+ age-group who were planning to move were asked what would entice them to move sooner, their responses were as follows:

- House is old, 8.1%
- Want only one floor, 7.1%
- Not enough storage, 3.8%
- Landlord, 2.8%
- Renting and prefer to own, 2.4%
- Traffic congestion, 2.4%
- Need to remodel, 1.9%
- Crime, 1.9%

Distance Moved

About one-third of recent movers who answered the survey had relocated 100 miles or more from their previous home. One-third had moved less than 10 miles from their previous home, and half had moved less than 20 miles away. Builders, developers, and other industry professionals should keep these percentages in mind when they are planning communities. The farther customers live from your development, the lower the probability that they will buy your home.

When householders move distances of 100+ miles they generally have a specific reason. The largest proportion said they wanted better weather or a better climate. Twelve percent said it was because they were retiring. Ten percent wanted to be closer to their friends, and 7% moved more than 100 miles because they had a property in a location that they liked. Only a small proportion of householders move to a regional lifestyle community just to enjoy its amenities and lifestyle. To capture customers from more than about a 20-mile radius, a community must be truly extraordinary.

Although there are differences in the number of miles moved among specific age-groups, the greatest difference is between householders younger than 75 and those who were 75 and older. Sixty-three percent of householders ages 75+ moved less than 20 miles compared with 52% or less of the age-groups younger than 75. Figure 6.10 shows distances moved by U.S. region.

For 45+, One Size Doesn't Fit All

The majority of boomers and silents are satisfied with where they live. Most moves result from

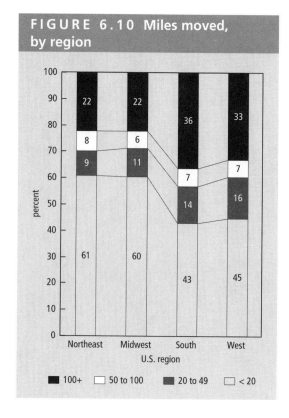

FIGURE 6.10 Miles moved, by region

people being attracted to something or someone, rather than from push factors discussed in Chapter 4. The more recently people have moved, the greater the probability they will move again. Conversely, the longer they have lived in their current residence, the lower the probability they will move. Thus, building a new community in an area in which many home owners have lived in established neighborhoods of ranch-style homes for more than 20 years may not yield the desired sales pace.

Keep in mind that these householders have a wide array of reasons for moving. Therefore, there is no one-size-fits-all home or community for Americans ages 45+. Although some predictable differences between householders result from age, income, marital status, and other attributes, most householders have many reasons for moving, and the impact of each may be greater or lesser, depending on the attributes of the homes and communities they encounter in their home searches.

PART III

WHAT THEY WANT

Preferences for Location, Home Site, and Community

Most people ages 45+ prefer to live in an outlying suburban area or a small town, and most prefer to live on or have a view of green space. They are most likely to want to live within walking distance of places they like to exercise, whether those are bicycle paths, walking trails, fitness centers, or the grocery store. In addition, they want to be a short drive from their church, medical services, retail shops, and restaurants.

People in the 55+ age-group who are planning to move to active adult communities are more likely to prefer a perimeter fence or guard-gated community than householders ages 55+ who are planning to move to all-age communities. They also are more likely to prefer master-planned communities than householders ages 55+ who prefer to live in all-age communities.

This chapter reviews the places where 45+ middle-Americans prefer to move and details preferences by age-group, income, current home value, and whether they plan to live in an active adult or an all-age community.

Ideal Location

Forty-four percent of middle-American householders ages 45+ said they preferred to live in an outlying suburban area, which was defined in the survey as at the outer edge of developed areas of a city, with comparatively more single-family homes and neighborhoods and fewer commercial areas (fig. 7.1). Most preferred either a rural area or small town. Only a small minority preferred a close-in suburban or urban location.

Home Site Characteristics

More specifically, householders ages 45+ were asked to choose their preferred home site from a list of options. Respondents were told that being "on the water or on the beach" would be more expensive than having a view of either of them. The questionnaire asked them to be realistic in their replies and to select the area that would best represent the site for their next home. Essentially equal proportions chose to be on a green space, have a view of green spaces or parks, or have a mountain view (fig. 7.2). Thirteen percent said they wanted a view of a lake, pond, river, or stream. Only 1% preferred to be on a golf course or to have a view of a golf course.

Although golf courses have long been thought to be the premier amenity in lifestyle communities, studies of golf course communities reveal that only 20% of residents of golf course communities are actually golfers. Residents are more likely to be attracted to golf courses as amenities because of the atmosphere they create—plenty of green space, nice views, and a social gathering place. Thus, developers can create the desirable ambience of a golf course community without the expense of adding a golf course.

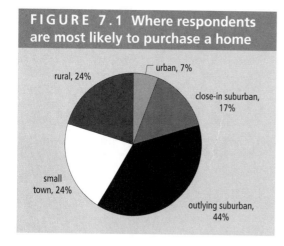

FIGURE 7.1 Where respondents are most likely to purchase a home

- urban, 7%
- close-in suburban, 17%
- outlying suburban, 44%
- small town, 24%
- rural, 24%

Walkable Communities

When ProMatura conducts focus groups with householders ages 45+, what these participants say more frequently today than in the past is that they want to be able to walk to something, rather than get in their cars to do everything. Many have said they would like to be able to walk to get a cup of coffee, to eat out, to have a drink at a neighborhood pub, or even to pick up a few items at a neighborhood store. They summarize their desires by saying, "I just want someplace nice that I can walk to."

The greatest proportion (18%) of middle-American householders ages 45+ said they would like to be within walking distance of a bicycling area. Another 17% would like to walk to a hiking area; 16% would like to walk to a grocery store; and 14% would like to be able to walk to a fitness center.

Driving Distances

Although many developers who target the empty-nester market think they need to locate their projects near a hospital or medical facility, survey results show that only 39% wanted to be within a 5- to 10-minute drive of a major medical facility. However, 72% of respondents wanted to be within a 10-minute drive of a grocery store. Still, 56% of survey respondents wanted to be no more than 10 minutes away from an emergency medical facility. In addition, 44% wanted to be within a 10-minute drive of a church of their denomination, and 37%–42% wanted to be within 10 minutes of retail shops, restaurants, areas for bicycling, and physicians' offices.

More than half wanted to be within an 11- to 30-minute drive of: physician's offices (58%), restaurants (52%), movie theaters (52%), major medical facilities (52%), and retail shops (51%). Between 40% and 48% wanted to be within an 11- to 30-minute drive of a major airport, performing arts center, emergency medical facilities, and a college or university.

Community Preference

A significantly greater proportion of middle-American householders ages 45+ planning to move said they would prefer a conventional neighborhood (67%) over a master-planned commu-

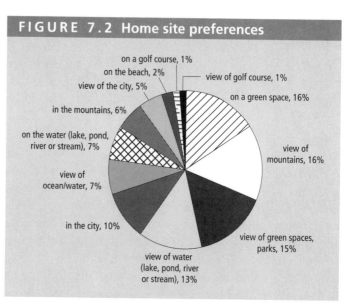

FIGURE 7.2 Home site preferences

- on a golf course, 1%
- on the beach, 2%
- view of the city, 5%
- in the mountains, 6%
- on the water (lake, pond, river or stream), 7%
- view of ocean/water, 7%
- in the city, 10%
- view of water (lake, pond, river or stream), 13%
- view of green spaces, parks, 15%
- view of mountains, 16%
- on a green space, 16%
- view of golf course, 1%

nity (13%). The description of a master-planned community that was provided to the survey participants is the same definition used by the Urban Land Institute: a suburban plan that includes homes; commercial establishments; and work, educational, and community facilities; with a logical and convenient relationship to one another. In the survey, the conventional neighborhood was described as "a neighborhood not in a master-planned community." Those who did not select either of these two choices selected other alternatives such as a gated community or an age-qualified community.

The proportion that planned to move either to a conventional neighborhood or to a master-planned community decreased with age (fig. 7.3). Fifty-two percent of householders ages 45–54 who planned to move said they were going to a conventional neighborhood, compared with 10% of the 75+ age-group. Some didn't select any of the choices offered, which suggests that they did not know where they might move. Among householders with higher incomes, the propor-

tion that planned to move to a master-planned community was higher (fig. 7.4).

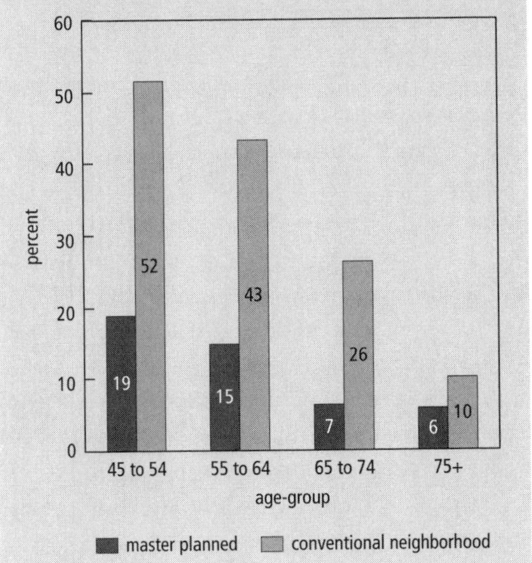

When survey responses were examined according to the household's current home values, the proportion of respondents who planned to move to a master-planned community followed essentially the same pattern as for household income.

Householders planning to move to an active adult community were more likely to say they were moving to a master-planned community (17%) compared with those planning to move to an all-age community (10%). Householders moving to an all-age community were more likely to say they were moving to a conventional neighborhood (46%) than those moving to an active adult community (24%). Respondents who planned to move to active adult communities were more likely to seek communities with amenities and convenient services than households planning to move to conventional neighborhoods.

Perimeter Fence, Gate, or Guard-Gated Community

Despite the long-held myth that as people grow older they become more concerned about safety and security, the proportion of householders who planned to move to communities with perimeter fences, gates, or guarded gates decreased as age increased. Householders ages 45–54 were almost seven times more likely to move to a community with a perimeter fence (7%) than householders ages 75+ (1%). This same pattern holds for those seeking gated or guarded and gated communities. More than two decades of research has demonstrated that younger householders are the most concerned about security. Householders in their 80s are more likely to have lived in their neighborhoods longer, know them better, and have the experience of time that helps them understand the probability of their security being breached in their homes and/or neighborhoods.

FIGURE 7.3 Community or neighborhood preference, by age-group

percent — age-group: 45 to 54: master planned 19, conventional neighborhood 52; 55 to 64: master planned 15, conventional neighborhood 43; 65 to 74: master planned 7, conventional neighborhood 26; 75+: master planned 6, conventional neighborhood 10.

■ master planned   ▨ conventional neighborhood

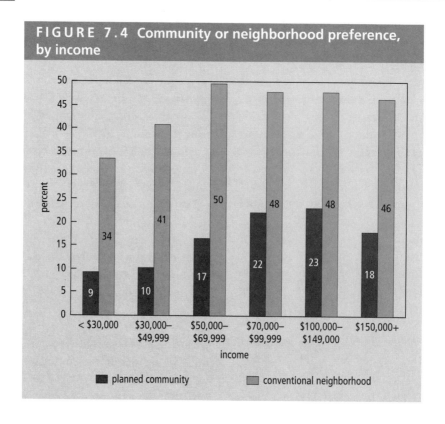

FIGURE 7.4 Community or neighborhood preference, by income

planned community   *conventional neighborhood*

Although the proportion of householders who plan to move to a community with a perimeter fence, gate, or guard at the gate generally increases with household income and home value, these households are still a minority. Only 17% of respondents with $150,000 or more in income planned to move to gated communities.

Householders moving to active adult communities were more interested in a perimeter fence (10%) and a guard at the gate (11%) than those moving to all-age communities. However, those moving to all-age communities were more likely to say they wanted a gated community (11%) than those moving to active adult communities (6%).

Householders living in the South and the West were significantly more likely to say they wanted to live in a community with a perimeter fence, gate, or guard at the gate than those in the Northeast and Midwest. Midwest householders were the least likely to say they wanted a perimeter fence (2%), gated community (4%), or guard at the gate (2%) (fig. 7.5).

Golf or Resort-Style Community

Prospective home buyers prefer to live in a community that is more like a small hometown than like a resort. Even when asked to think of the nicest resort they have ever visited and to think of that resort as being a place where their friends and neighbors also live year round, the vast majority still say they prefer to live in a small, hometown community.

Among the households planning to move, nearly equal proportions said they wanted to relocate to a golf course community (4.5%) or a resort-style community (4.6%), but preferences differed by age-group, and the proportion of householders likely to choose either a golf course or resort-style community decreased as age increased. Some 5%–7% of people ages 45–54 would choose a golf course or resort-style community compared with 1% of the householders 75+.

As income increased, so did the proportion that preferred to move to a golf course or resort-style community. Fourteen percent of the households with $150,000+ in annual household income said they prefer a golf course community, and 9% said they prefer a resort-style community (fig. 7.6).

Householders exhibit a similar pattern when their preferences for a golf course or resort-style community are compared to their home values. Householders with higher home values prefer either type of high-amenity community more than households with lower home values. Essentially equal proportions of households with homes valued at $400,000 or more prefer a golf course community as want a resort-style community (11%).

Householders who plan to move to an active adult community are more likely to prefer a golf

course community (8%) or a resort-style community (13%) than households who prefer an all-age community (3%).

Householders in the South and the West are more likely to choose a golf course or resort-style community than households in the Northeast or Midwest. Between 6% and 7% of householders in the South and West chose the communities with recreational amenities, compared with 2%–4% of the households in the Northeast and Midwest. Twice the percentage of Northeasterners (4%) wanted a resort-style community as wanted a golf course community (2%).

Dispelling Myths about 45+ Buyers

Many myths and assumptions about boomers and silents were dispelled in this chapter. First, the data indicate that of many desirable locations (green space, water view, beachfront, mountain view, or on a golf course) the smallest proportion

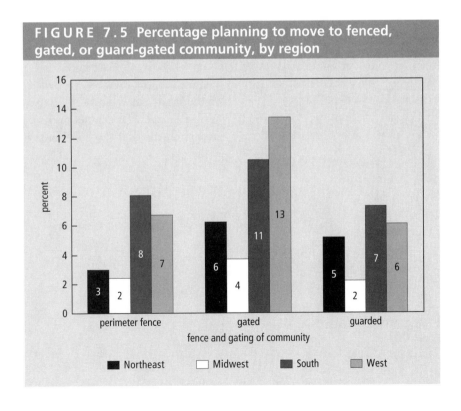

FIGURE 7.5 Percentage planning to move to fenced, gated, or guard-gated community, by region

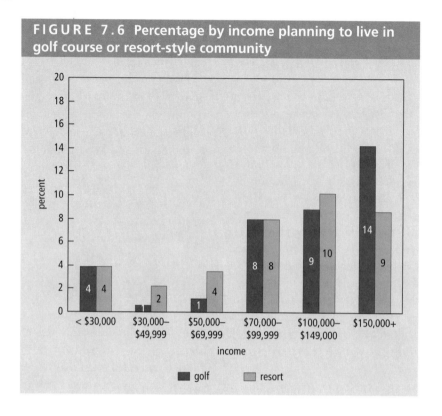

FIGURE 7.6 Percentage by income planning to live in golf course or resort-style community

of respondents chose a view of a golf course or to live on a golf course.

Also contrary to common wisdom, younger households are more concerned about security and safety than are older households.

Finally, buyers ages 55+ are more concerned about being close to fun and interesting places that they can walk to than they are about being close to a major medical facility.

Yet this chapter also demonstrates that boomers and silents aren't the same everywhere. Households in the South and West are at least twice as likely to want recreational amenities in their communities as are households in the Northeast or Midwest.

8

Preferences for the Style and Appearance of a Community

The style and appearance of a community create an ambiance, set a certain tone, and establish a mood. A community can be relaxing and energetic at the same time. Although outward appearance is important, and is a factor in whether a community will appeal to a buyer, of more importance is how a community's design and features will bring people together. What home seekers encounter in the first few moments of a visit can compel them to buy, stimulate their curiosity, or ensure that they will never return. The community entrance, scenery on the drive to the sales center, sales staff, and the first face—ambivalent or aloof, frowning or happy, abrupt or gracious—can make for either a one-time visit or the journey of a lifetime.

Gathering places, sidewalks, shops, clubhouses, and sales centers must communicate excitement and expectancy even when empty. The community square must be a place, not a pass-through. The lobby or entry must be welcoming.

This chapter focuses on the tangibles of a community. It will examine preferences for home and community style, landscaping, entrance, front yards, open spaces, trees, traditional neighborhood design versus suburban garage-in-front design, sidewalks, security, maintenance, and views. The design of the site plan, landscape and buildings, the way they interrelate, the attention to detail, relationship of one public area to another public area, and the interface of public spaces and residential areas will suggest either a vibrant com-

You have one chance to make a first impression. Take all the negatives and turn them into positives or at least neutrals. The community must be clean, crisp, lively, and inviting. We must remember that we are selling a community and a lifestyle, and we must first focus on this. Once we sell our clients on the community, then we begin to focus on which home or home site will best suit them.

—Douglas Diez
Developer and Resident
Pelican Point Properties, LLC
Gonzales, Louisiana

munity or just lovely window dressing. A vibrant community is what captures sales.

A community with a beautiful and grand entrance and lavish sales center with row after row of houses and a clubhouse reached only by getting in a car exemplifies superficial window dressing with little attention to creating an attractive site plan. In contrast, a community with a carefully planned series of walkways, gardens, green spaces, gathering places, and any number of interesting ways to ambulate through it and casually encounter neighbors offers a community with greater opportunity for a more vibrant lifestyle.

> *The most important thing we did when planning a community was to gather key constituents (construction, land development, finance, marketing, management, sales) to discuss "a day in the life" of our future residents beginning on the morning of a weekday and continuing through the weekend, 24/7. We then planned our community around who they are and what they did (or ideally would like to do). Interestingly, it affected what we build, how we build it, and what lifestyle we delivered in the community. In the absence of this exercise, we would have missed several key elements that would have made our venture more difficult.*
>
> —J. Scott Glaus
> Vice President, Strategic Marketing
> Centex Homes
> Chantilly, Virginia

The Entrance Says "Hello"

Whether your entrance is grand or simple, guarded or wide open, it should say "Hello."

An attractive, eye-catching entrance turns heads. It invites potential prospects to explore, rather than just drive by. If customers then find value in the product inside, they become prospects for a sale. Although more than two-thirds of the 45+ middle-American households said they were indifferent to an "upscale entrance" to a community or that it was unimportant, one-third said an upscale entrance matters.

Householders ages 55+ were more likely than the 45–54 age-group (37% or more compared with 29%) to think that an upscale entrance is important.

The preference for an upscale entrance (fig. 8.1) increases with income. About a quarter of house-holds with incomes of $30,000–$70,000 wanted an upscale entrance. Thirty-four percent of those with annual incomes of $70,000–$99,999 wanted one, and 48% with incomes of $150,000 wanted an upscale entrance.

Two-thirds of the consumers looking for a gated community or a community with a guard at the gate said that an upscale entrance was important. Thirty-nine percent interested in a master-planned community said that an upscale entrance was important. Only 14% of those thinking of moving to a conventional neighborhood thought so.

Half of the householders who planned to move to an active adult community said that an upscale entrance was important (fig. 8.2). In contrast, 35% of householders who preferred "either" an active adult or an all-age community and only 23% of the householders who planned to move to an all-age community preferred an upscale entrance.

More than half of the households (53%) considering an attached home said that an upscale entrance was important, very important, or essential. As readers will learn in Chapter 10, a sizable proportion of the householders who want an attached home are thinking of moving to an active adult community whereas a large portion of householders who want a detached home are planning to move to an all-age community. Forty percent of the householders planning to buy a home in a multifamily, multistructure community wanted an upscale entrance for that community. Apparently, people moving to a place for a stronger experience of community want the entrance to make a statement about the community.

About half of the householders who planned to spend more than $300,000 for a home wanted an upscale entrance at their community (fig. 8.3). Slightly less than one-fourth of the householders who said the highest price they would pay for a new home was under $150,000, and close to 30% of those who estimated they would pay up to $299,999 said an upscale entrance was important.

Almost 50% of the households in ZIP Code Region 2 (Mid-Atlantic) and ZIP Code Region 3

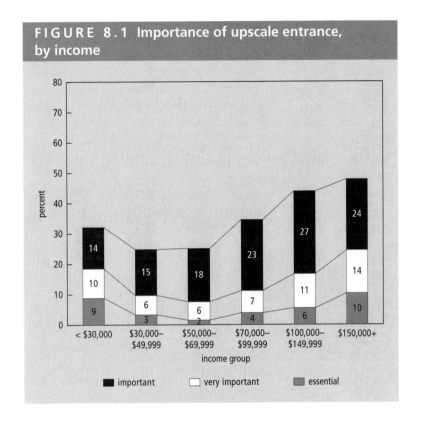

FIGURE 8.1 Importance of upscale entrance, by income

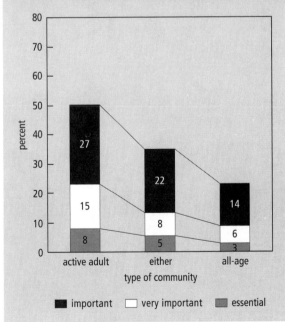

FIGURE 8.2 Importance of upscale entrance, by preference for active adult or all-age community

(Southeast) said an upscale entrance was important (fig. 8.4). Less than one-fourth of householders planning to move in ZIP Code Region 4 (Ohio Valley) and ZIP Code Region 5 (Northern Plains) said an upscale entrance was important.

Landscaping

Upscale landscaping can be of many different styles—from formal, with decorative, pruned hedges and stately gardens, to natural, with native plants, wildflowers, and grasses. In either case, these landscapes incorporate a variety of plants, both large and small, as well as water features, such as fountains, and hardscapes, such as sculptures and outdoor living spaces.

As with an upscale entrance, the desire for upscale landscaping increased with respondents' incomes. Two-thirds of householders with $150,000+ in annual household income

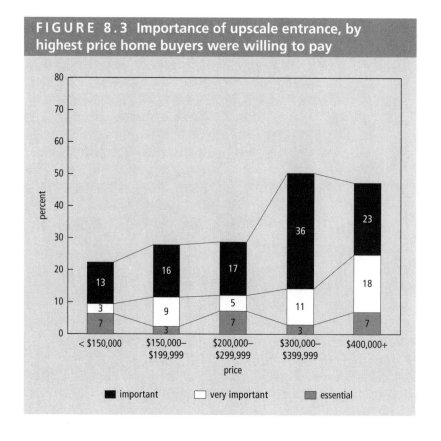

FIGURE 8.3 Importance of upscale entrance, by highest price home buyers were willing to pay

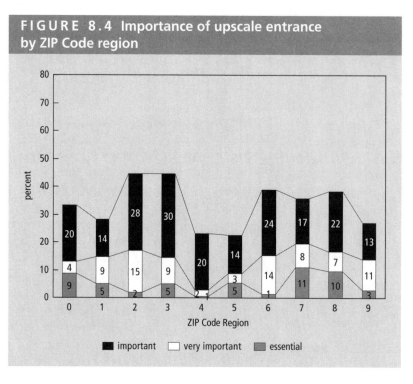

FIGURE 8.4 Importance of upscale entrance by ZIP Code region

said that upscale landscaping was essential, but none of the households with annual incomes of less than $50,000 said that upscale landscaping was essential (fig. 8.5).

Likewise, as the price that householders ages 45+ are planning to pay for their homes rises, so does their expectation for upscale landscaping. Sixty-four percent of the householders planning to spend $400,000 or more on their homes said they wanted upscale landscaping (fig. 8.6).

Moreover, beautiful landscaping appears to be of greater concern to householders ages 45+ than an upscale entrance. Although one-third of them said an upscale entrance was important, almost half (47%) said upscale landscaping was important. More than half of those in the 55–64, 65–74, and 75+ age-groups said upscale landscaping was important.

Although the proportion of householders who wanted to live in a community with a guard at the entrance was small, this group probably will keep landscaping companies in business. Eighty-five percent of the households planning to move to a community with a guard at the gate believed that upscale landscaping was important (fig. 8.7). In contrast, 38% who were thinking of moving to a conventional neighborhood said upscale landscaping was important.

Clearly, the householders planning to move to an active adult community want an aesthetically pleasing environment. Seventy-two percent of them, compared with one-third of the householders planning to move to an all-age community, said upscale landscaping was important (fig. 8.8).

Preferences regarding landscaping differed by region. In the Mid-Atlantic (ZIP Code Region 2), 61% of householders said upscale landscaping was important (fig. 8.9). However, none of the households in the Ohio Valley (ZIP

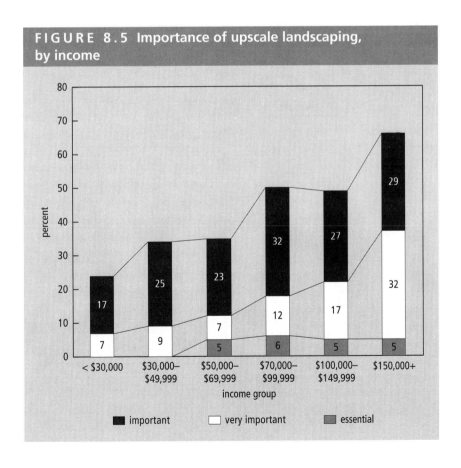

FIGURE 8.5 Importance of upscale landscaping, by income

percent / income group

important   very important   essential

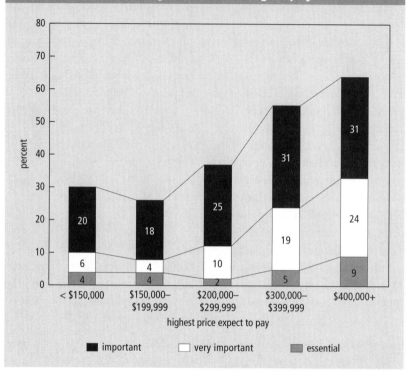

FIGURE 8.6 Importance of upscale landscaping, by highest price home buyers were willing to pay

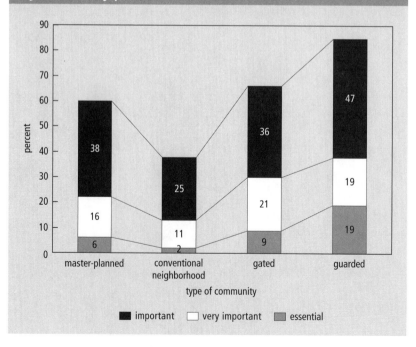

FIGURE 8.7 Importance of upscale landscaping, by community preference

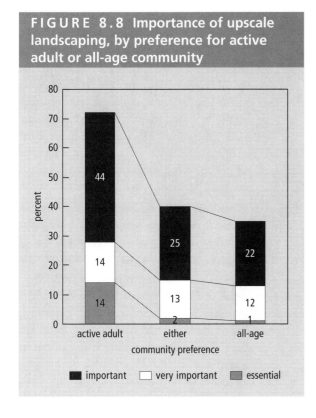

FIGURE 8.8 Importance of upscale landscaping, by preference for active adult or all-age community

Code Region 4), Midwest (ZIP Code Region 6), or Mid-South (ZIP Code Region 7) said upscale landscaping was essential. Still, roughly 40% of the householders in these regions said upscale landscaping was important.

Front Yard Upkeep

Builders should not misinterpret the desire for upscale landscaping as a desire on the buyers' part to be freed from yard maintenance. In fact, a majority of 45+ households do not want front yard upkeep. Only about one-third of the market wants to be freed from yard maintenance, so don't send the message that people who enjoy yard work or gardening need not move to your communities.

Slightly more than half of 45+ householders said they might consider a service to keep up their front yards, and about one-

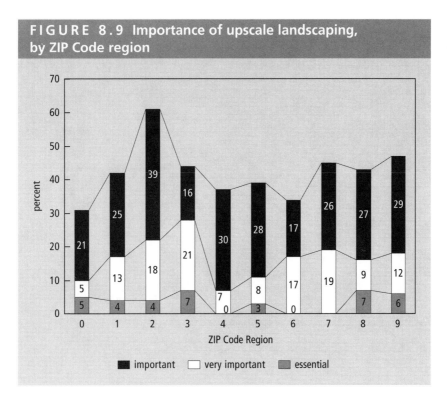

FIGURE 8.9 Importance of upscale landscaping, by ZIP Code region

third of middle-American householders ages 45+ planning to move believed that having a service to take care of the front yards of the residences in their community was important. A significant proportion of householders move to get away from the constant upkeep of a home and yard, and the proportion that believed it's important to have a service to keep front yards in good shape increased with age. Twenty-eight percent of the 45–54 age-group surveyed said that front yard upkeep was important, compared with 49% of householders who were ages 75+.

In contrast, householders with lower incomes were not interested in spending their money on front yard upkeep. Only 17% of householders with less than $30,000 in annual household income said front yard upkeep is important; none said it is very important or essential. Between 29% and 36% of householders with annual incomes greater than $30,000 said front yard upkeep is important, very important, or essential.

Householders who move to a guarded, gated community are much less inclined to want to spend their time in their yards than household-

ers who prefer a conventional neighborhood. Twenty-three percent of households thinking about moving to a conventional neighborhood said that front yard upkeep is an important service. Forty-five percent of the householders who said they were likely to move to a master-planned community said that front yard upkeep services are important, as did 52% of householders planning to move to a gated community and 66% of those planning to move to a community with a guard at the gate (fig. 8.10).

The majority of householders planning to move to an active adult community (54%) apparently aren't moving there so they can spend their time keeping up their yards (fig. 8.11). Twenty percent of the householders moving to an active adult community considered upkeep of the front yard an essential service. In stark contrast, only 2% of the households moving to an all-age community considered services for their front yards essential.

Fifty-seven percent of householders likely to purchase an attached home said that front yard upkeep services were important. Again, the reason

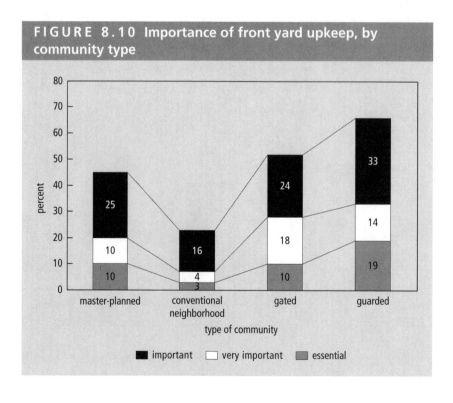

FIGURE 8.10 Importance of front yard upkeep, by community type

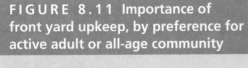

FIGURE 8.11 Importance of front yard upkeep, by preference for active adult or all-age community

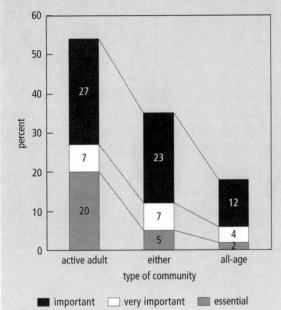

Diverse Home Styles

Forget about using a cookie-cutter approach to designing for the boomer market. Well over half of 45+ middle-American householders planning to move said that diversity of home styles in a community is important, and only 21% said that a variety of home styles is unimportant. A slightly smaller proportion of 75+ householders than younger householders think variation in home styles is essential or very important.

Figures 8.13 and 8.14 show the breakdown by income and type of community desired.

The most important lesson in the 55+ market has been that you must allow for "personalization" of the home and not build out the inventory too far. The investment is well worth it in benefits of additional option dollars and referrals from happy home owners.

—ED SCANNAPIECO
Senior Vice President Sales and Marketing
Abbot Homes
Boston, Massachusetts

may have more to do with ensuring that neighbors' yards are maintained than with desire to rid themselves of yard maintenance. Half as many householders moving either to a single-story or to a two-story detached home, compared with those moving to an attached home, said upkeep services for front yards are important.

Although minor differences exist between the groups based on the most they said they would pay for their homes, the data show no specific link between desired home price and whether buyers wanted lawn and landscaping services. Householders planning to spend less on their homes may include a greater proportion of buyers who are planning to purchase an attached home, and it may include householders in the 75+ age-group who are more likely to want yard services.

The combined percentage by ZIP Code region of survey respondents that said that front yard upkeep services were essential, very important, or important ranged from 14% in Region 5 to 46% in ZIP Code Region 0 (fig. 8.12).

Builders often ask whether the active adult market can "read" floor plans and if they have a good idea of what they want in a home. Generally, the 55+ population has had significant experience living in a house—slightly more than those younger than 55. Figure 8.15 suggests that these buyers have a good idea of what they want. A significantly greater proportion of householders planning to move to an active adult community than those in the other two groups wants to see an interesting streetscape. Moreover, 80% of the householders considering purchasing an attached home said a variety of home styles in the community is important.

Among all of the categories of 45+ buyers surveyed, only those considering a home in the

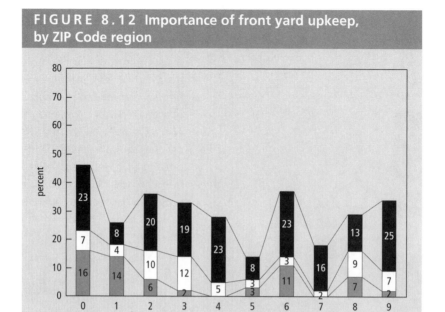

FIGURE 8.12 Importance of front yard upkeep, by ZIP Code region

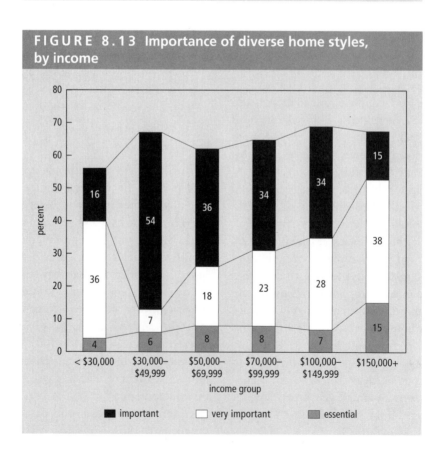

FIGURE 8.13 Importance of diverse home styles, by income

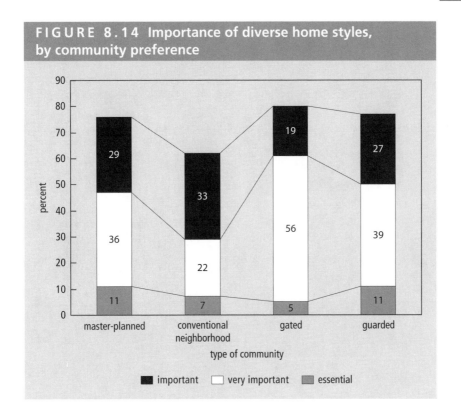

FIGURE 8.14 Importance of diverse home styles, by community preference

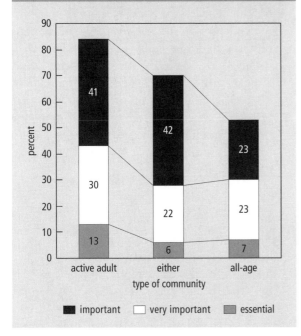

FIGURE 8.15 Importance of diverse home styles, by preference for active adult or all-age community

$150,000–$199,999 price range demonstrated a relatively weak preference for diversity in home styles. Even so, 41% of them said that diversity was important, very important, or essential.

Apparently though, householders in the northern tier of the United States (ZIP Code Regions 4 and 5) do not feel as strongly about variation in home styles as those in the Eastern, Mid-South, and Western regions (fig. 8.16). Still, more than half of the households in the Northern Plains and Ohio Valley said that diversity in home styles is important.

Green and Open Spaces

For years participants in surveys, focus groups, and other research have indicated that a view is important. In our survey, two-thirds of respondents said having a great view was important. Many other buyers probably

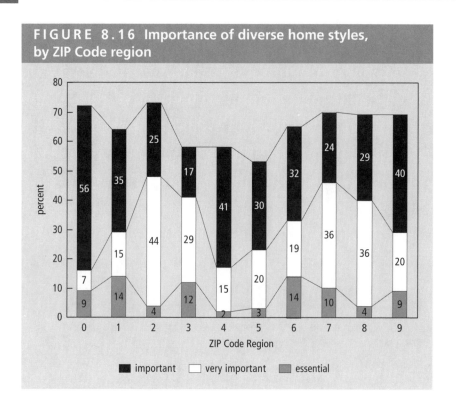

FIGURE 8.16 Importance of diverse home styles, by ZIP Code region

would be pleased just to have a nice view, such as one of a mature tree. Seventy percent of middle-American householders ages 45+ said that green and open spaces in a community are important. Most respondents aren't dreaming of an expansive view, but rather a pleasant space that might have a tree, a water feature, or a landscaped area with seasonal color.

Therefore, create views even if they are modest, and plan your communities to help buyers focus on their beauty. A beautiful view can be as simple as a lovely flower to behold. A trip to Japan helped reinforce how easily this might be accomplished. On this trip, our host took us to a restaurant where he seated us with our backs to the wall so that we might look out into the restaurant, as many of us like to do. A few moments after we had been seated, however, he apologized and said that he was sorry that he had taken the seat with the view, and offered for us to trade places. Knowing a wall was behind us, we wondered what he meant, but turned around and looked

anyway. In a recess in the wall there was a small orchid in bloom.

With the sole exception of those willing to pay $150,000 or less for a home, more than half of respondents—by income group, type of community desired, home style, price respondents were willing to pay, and ZIP Code region—considered green or open space important, very important, or essential. The proportion of householders that believe that open and green spaces are important increases steadily with the price they are willing to pay for a home (fig. 8.17). Regional data are detailed in figure 8.18.

Mature Trees

Along with their preference for green space, 45+ households also expressed a strong desire to have mature trees in their surroundings. The proportion that believes mature trees are important increases with household income (fig. 8.19). Look for these buyers to patronize builders and

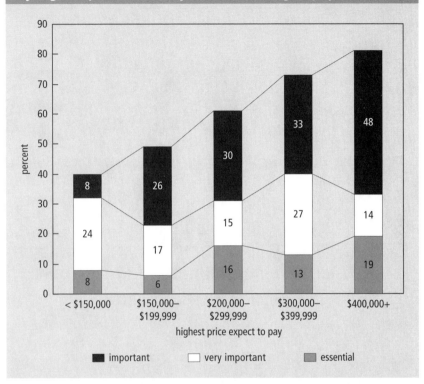

FIGURE 8.17 Importance of green or open space, by highest price home buyers were willing to pay

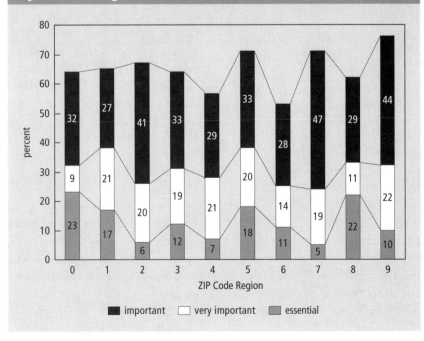

FIGURE 8.18 Importance of green or open space, by ZIP Code region

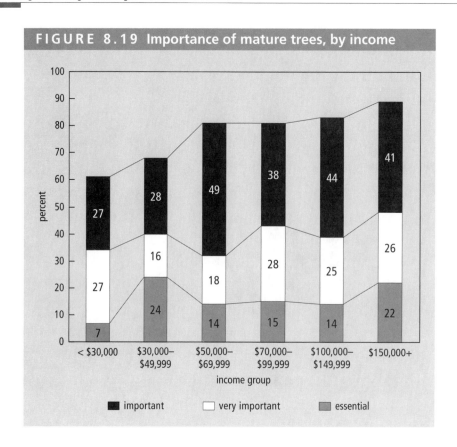

FIGURE 8.19 Importance of mature trees, by income

developers who learn to keep the trees in developments. The importance respondents placed on mature trees by ZIP Code region is shown in figure 8.20.

Neighborhood Style and Home Design

Almost equal proportions of 45+ middle-American householders agreed that they would prefer a home in a community with a traditional neighborhood design (garage in back of the home) and a suburban-style community with the driveway and garage on the front of the home, but less than 30% expressed one preference or the other. No statistically significant differences arose in the householders' preferences for the two styles by the respondents' age or their annual household income.

In the survey, traditional neighborhood design was described as "the new design of communities that reflects how small towns used to be designed (big front porches that are closer to the street, sidewalks, cars parked behind houses, park areas, and a town center)." The suburban style was described as "the style of community with garages on the front of homes and homes set farther back from the street" than in traditional neighborhood design. Respondents were shown pictures of both types of neighborhood design.

More households planning to move to a master-planned community or gated community than those planning to move to a guarded and gated community or a conventional neighborhood strongly preferred traditional neighborhood design (garage in back) than suburban-style neighborhood design (garage in front). Still, less than 20% expressed a preference. Among those interested in moving to a conventional neighbor-

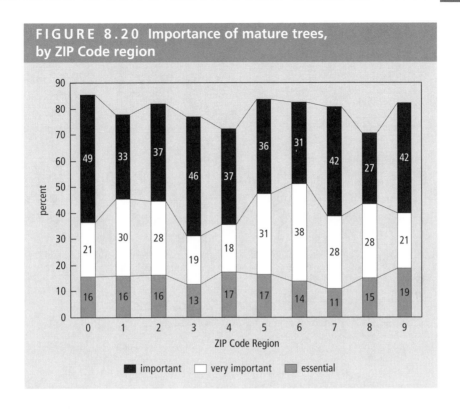

FIGURE 8.20 Importance of mature trees, by ZIP Code region

Sidewalks

hood, equal proportions of householders (9%) strongly preferred either style of house.

Among those who planned to move to an active adult community or to either an active adult or all-age community, equal proportions said they strongly preferred a traditional neighborhood and a suburban-style neighborhood, but again, the percentages were small (16%). Among householders who planned to move to an all-age community, a greater proportion strongly preferred suburban-style design (10%) over traditional neighborhood design (6%).

Finally, looking at survey responses about community design by ZIP Code region, the strongest positive reaction to traditional neighborhood design was in ZIP Code Region 8, where some 22% expressed that preference compared with 3% who said they preferred the suburban-style design (fig. 8.21). In contrast, in ZIP Code Region 6, 9% preferred suburban-style and 3% preferred traditional neighborhood design.

Sidewalks

Even in areas with only a modest amount of traffic, 45+ buyers as a group said they want sidewalks. Some 70% of them said that sidewalks are important in their communities. This preference for sidewalks held across age, income group, and highest price respondents were willing to pay for a new home. However, more 45+ buyers who were planning to move to an attached home or a multi-family residence than those planning to move to a detached residence considered sidewalks essential. In addition, householders who preferred conventional neighborhoods were much less likely to say sidewalks were important than those considering master-planned, gated, or gated and guarded communities (fig. 8.22). Moreover, those respondents planning to move to all-age communities were much less likely to say that sidewalks were essential. Only 21% of them said so, compared with 42% of those who planned to move to active adult communities (fig. 8.23).

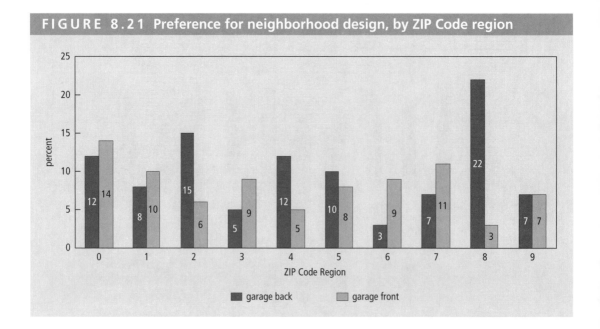

FIGURE 8.21 Preference for neighborhood design, by ZIP Code region

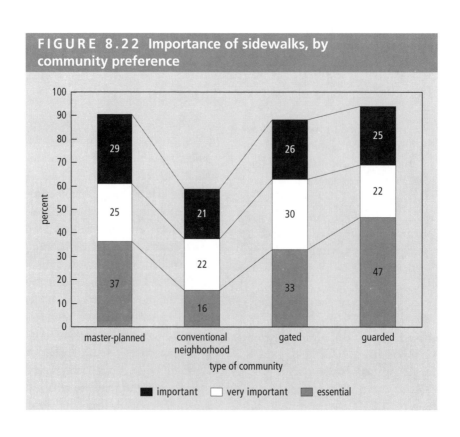

FIGURE 8.22 Importance of sidewalks, by community preference

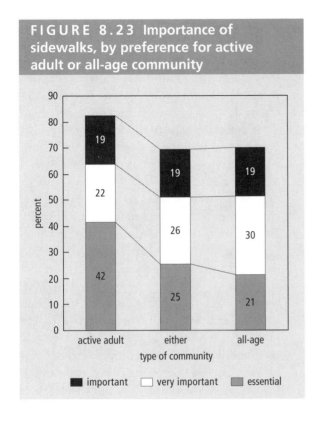

FIGURE 8.23 Importance of sidewalks, by preference for active adult or all-age community

Regionally, householders in the Western Mountain states (ZIP Code Region 8) demonstrated the strongest preference for having sidewalks (fig. 8.24). Recall that this group also was among the strongest supporters of traditional neighborhood design.

A Lock-and-Leave Home

Not surprisingly, 96% of householders planning to move want to be able to lock and leave their homes. More than half of the householders in ZIP Code Region 8, the Western Mountain states, said being able to lock and leave their homes was absolutely necessary. These states include a high proportion of second, vacation, and retirement homes.

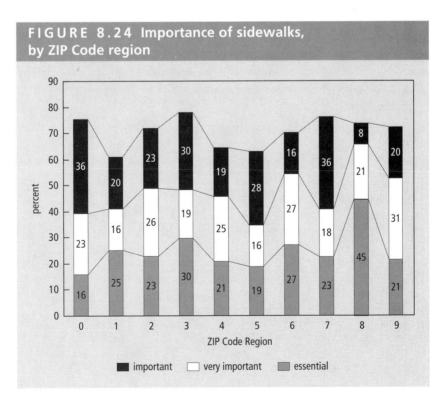

FIGURE 8.24 Importance of sidewalks, by ZIP Code region

Outdoors Matter to Buyers

As housing densities increase and lot sizes decrease, the appearance of outdoor spaces becomes more important. In fact, outdoor spaces are as important as interior spaces to 45+ housing consumers. Moreover, although some of these buyers may be willing to trade an upscale entrance for another feature, keep in mind that an entrance sets the tone for an entire community. Although an entrance doesn't have to be lavish, it should be designed as a community asset—one that will let prospective buyers know they are home.

Also remember that landscaping is important to almost half of the householders planning to move. Even if respondents say it is not, know that they will choose a home in a community with landscaping over a home in an area that doesn't have it. Landscaping can create a view in areas of the country that don't have enough existing natural views.

As important for some segments of the 45+ market is the opportunity to create their own natural outdoor spaces. Developers and builders may disenchant more than half of the market by suggesting to these buyers that they won't have an opportunity to get their fingers in the dirt, even if some boomers and silents want to rid themselves of yard work. So remember that the importance of lawn and landscaping services varies among different groups of 45+ buyers.

Just as buyers want interesting landscapes to look at, they also want diverse streetscapes. So, by all means, build your communities around a theme, but don't make them monotonous. For every community, plan as much green space as possible and try to give as many homes as possible a view of the green or open space even if it is only a small area. Two ways to accomplish this are to incorporate green throughout the community by constructing many small parks, rather than one large one, and to save trees, which are among the most coveted amenities. In fact, developers probably will be able to build smaller clubhouses if they keep more trees and establish walking trails meandering through them. Keep in mind that lots with trees can yield higher premiums.

Sidewalks, like parks and trails, can be a significant catalyst in creating community and in helping to sell homes. If community members find that getting from one place to another in the community is easy and interesting and if there are places for them to walk to, they will get out and meet their neighbors. The more lively your community, as evidenced by people being out and enjoying it, the better your homes will sell.

9

Community Amenities

What community amenities do boomers want? The answer depends, among other factors, on whether they are planning to move, their age, what they are willing to spend for a home, and where they reside now.

Data Collection

The survey explored the preferred amenities of 1,145 middle-Americans 45+ years old who either were planning to move within five years or who had relocated in the two years prior to being surveyed. Respondents were asked to rate the importance of 24 common amenities as "essential," "very important," "important," "indifferent," "unimportant," or "very unimportant." They chose one rating for each of the 24 amenities. In this chapter, where respondents are described as rating an amenity as important+, the rating refers to the combined percentage of respondents that rated an amenity as "important," "very important," or "essential."

Walk This Way

Among those who planned to move within five years and those who had relocated in the two years prior to the survey, walking trails drew the strongest favorable response from among the two dozen choices of community amenities the survey asked about. Other amenities that comparatively larger percentages of respondents planning a move rated as important or higher included a neighborhood park, fitness center, indoor walking track, swimming pool, and clubhouse.

Respondents rated the importance of a basic clubhouse with space for meetings and social

It's all about the lifestyle at a Del Webb [a division of Pulte Homes] community. Although interest in many passive recreation activities still exists, demand for more active forms of lifestyle programming has emerged as a top priority among current residents and prospective home buyers. Even adventure programming, such as hiking, climbing, and river rafting are appealing to a much larger percentage of our residents than we have seen in past years. At Anthem Ranch by Del Webb, outside of Denver, Colorado, the lifestyle director has implemented skydiving, white-water rafting, hot-air ballooning, sports flying, hang gliding, and even parachuting adventures.

—**JUDY JULISON**
National Director of Lifestyle
Operations
Pulte Homes
Huntley, Illinois

The Cost of Amenities

Amenities don't grow like weeds—they don't just appear. Before they sell their first home, many builders and developers already have invested millions, if not hundreds of millions of dollars, in infrastructure such as streets, sewers, lights, curbs, and gutters, as well as community amenities. The more money invested, the higher the price of the homes. Home owner association fees pay for upkeep of common areas and community amenities. The price residents pay depends not only on the number and types of amenities, but also on the number of homes in their community. Fewer homes will mean larger maintenance fees. Therefore, larger communities may mean more amenities—and lower monthly fees—for home owners.

gatherings and a more elaborate clubhouse that also would include a theater, art studio, music rooms, and card and game rooms. Essentially equal percentages of survey respondents planning to move preferred the basic clubhouse as the clubhouse with more features. As with all of the survey questions, this one was not specific to a particular community.

> *Lifestyle is more than a generic clubhouse.*
>
> —GREG IRWIN
> Principal
> Irwin Pancake Architects
> Costa Mesa, California

Among the 766 survey respondents who had moved in the past 2 years, the top 3 amenities desired were walking trails, an outdoor pool, and an outdoor heated pool. Moreover, 4 items that were not among the top 10 preferred amenities of those planning to move ranked in the top 10 list of those who already had moved. These were an 18-hole golf course, a basic clubhouse, a full clubhouse, and an outdoor community barbeque area.

The survey results suggest amenities carried less weight in the post-movers' home location decision than they will in the future movers' choice of location. For example, although 29% of those planning to move rated walking trails as essential or very important, only 11% of those who moved in the past 2 years did.

Age Matters

Boomers are not a monolithic group, and the importance they assign to certain amenities varies somewhat by age. Walking trails are a top preference for those in the 45–54 and 55–64 age-groups. At the same time, more respondents in the latter group placed among their top amenities an indoor walking track (64% compared with 31%), a basic or full clubhouse, and a same-sex fitness center. Only 5% of those ages 45–54 said a same-sex fitness center was essential or very important.

Respondents ages 45–54 placed an outdoor pool or an outdoor heated pool among their top 3 choices, although a greater percentage of movers ages 55–64 rated an outdoor pool as important+. A slightly larger proportion of the latter group also rated an indoor pool as important+.

Among those who were 65–74 years old, 44% rated an indoor pool as at least important, and 19% said it was essential or very important. This age-group also rated a full clubhouse, basic clubhouse, and an 18-hole golf course among its top 10 essential or very important amenities.

Respondents ages 75+ were generally less interested in amenities than the other 3 groups. Compared with the other age-groups, fewer respondents in this age-group rated any of the amenity choices presented to them as essential or even very important. The 3 amenities rated most often as essential or very important and the percentage of respondents who rated them as such follows:

- full clubhouse (23%)
- walking trails (22%)
- basic clubhouse (21%)

Although 23% rated an indoor pool as important, only 4% said this amenity was either very important or essential.

Pet Precedence

Given Americans' overall fascination with their pets, the importance that some survey respondents placed on a dog park might be expected. This amenity ranked among the top 10 for those 45+ who were planning to move in the next 5 years, whereas only 5% or less of them said a children's pool and a children's playground were essential or very important.

Active Adults

The "active adult" label is well-founded: People who seek out these age-qualified communities are, indeed, active adults. In fact, adults ages 55+ planning to move to active adult communities and those planning to move to all-age communities provided markedly different responses to questions about amenities.

Regional and Income Preferences

Along with age and other factors, income and geographic region of households play a part in the amenities that 45+ adults find attractive. This section examines different preferences by the amount respondents planned to spend on their homes and by ZIP Code region for amenities rated as the top 5 among all survey participants: walking trails, outdoor heated swimming pool, outdoor swimming pool, indoor heated swimming pool, and dog park.

Walking Trails

Sixty percent of respondents in the Midwest were indifferent to or didn't think walking trails were important, but in the Mid-Atlantic 75% rated these trails as important+ (fig. 9.1). Nearly 20% of those in the Western Mountain region said walking trails were essential.

A significantly larger percentage of respondents (more than two-thirds) who expected

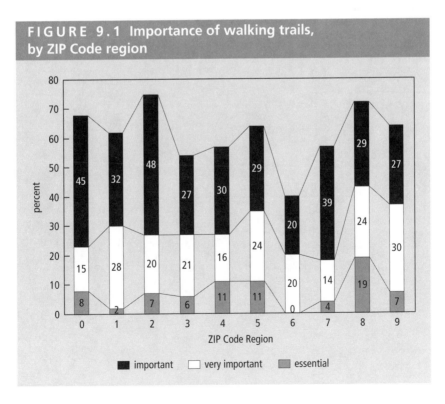

FIGURE 9.1 Importance of walking trails, by ZIP Code region

■ important □ very important ▨ essential

to pay more than $300,000 for their homes rated walking trails as important+ compared with respondents in other price ranges (fig. 9.2). About half of respondents in the $150,000–$199,999 and $200,000–$299,999 price ranges rated walking trails as important+.

Outdoor Heated Pool

Pacific Coast survey respondents (47%) were more likely than other survey respondents to rate an outdoor heated pool as important+. In comparison, 20% of Northern Plains respondents said that an outdoor heated pool was important+ (fig. 9.3).

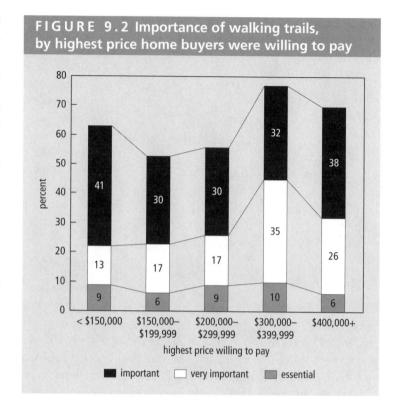

FIGURE 9.2 Importance of walking trails, by highest price home buyers were willing to pay

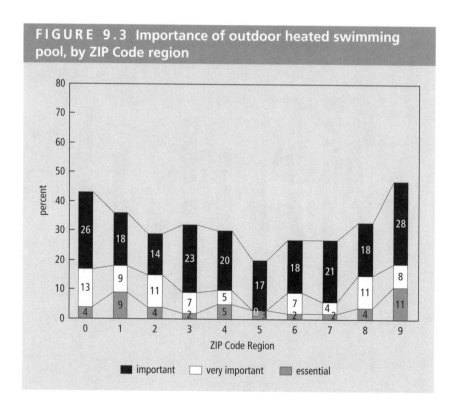

FIGURE 9.3 Importance of outdoor heated swimming pool, by ZIP Code region

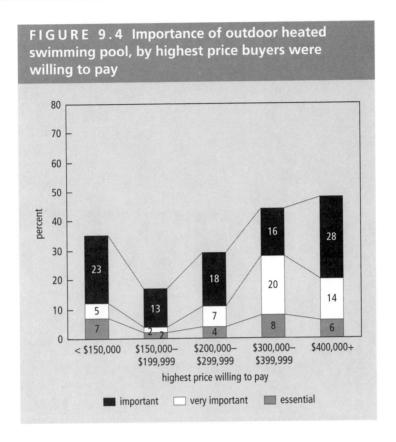

FIGURE 9.4 Importance of outdoor heated swimming pool, by highest price buyers were willing to pay

More than 44% of the householders planning to spend at least $300,000 for their homes rated an outdoor heated swimming pool as important+ compared with 17% of the householders planning to spend $150,000–$199,999 (fig. 9.4).

Outdoor Pool (Unheated)

As might be expected, survey respondents in colder regions were less likely to rate an outdoor pool as important. Although 43% of respondents in the Southern region that includes Texas and Louisiana said a swimming pool was important+, only 23% of those in the Northern Plains agreed (fig. 9.5).

After eliminating responses from those who wanted to spend less than $150,000 for a home (these respondents demonstrated a robust demand for amenities similar to that of the highest income groups); for all other groups, the importance of a swimming pool increased with the amount people were willing to spend for a home. At least 46% of the respondents willing to pay $300,000 or more for their homes wanted an outdoor swimming pool (fig. 9.6).

Indoor Heated Swimming Pool

Survey respondents in ZIP Code Regions 0 and 1 were the most interested in having an indoor swimming pool: At least 45% of these respondents said this amenity was important+. Only 27% of respondents in the Northern Plains said an indoor swimming pool was important+, the lowest proportion among households for all regions surveyed (fig. 9.7).

People willing to spend $300,000–$399,999 were most desirous of an indoor heated swimming pool (48%), but only 26% of those who wanted to spend $150,000–$199,999 said this amenity was important+ (fig. 9.8).

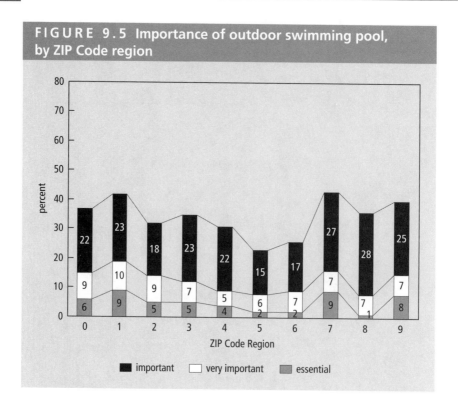

FIGURE 9.5 Importance of outdoor swimming pool, by ZIP Code region

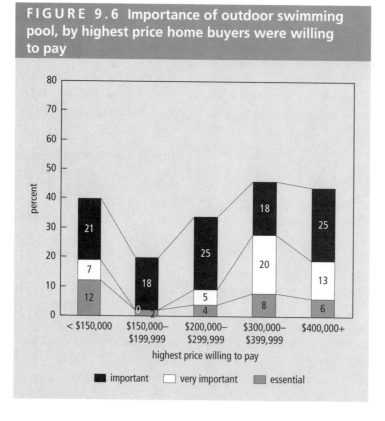

FIGURE 9.6 Importance of outdoor swimming pool, by highest price home buyers were willing to pay

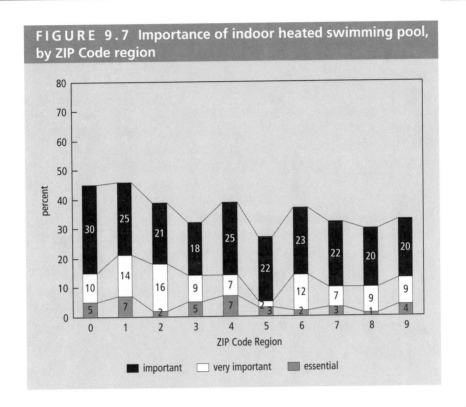

FIGURE 9.7 Importance of indoor heated swimming pool, by ZIP Code region

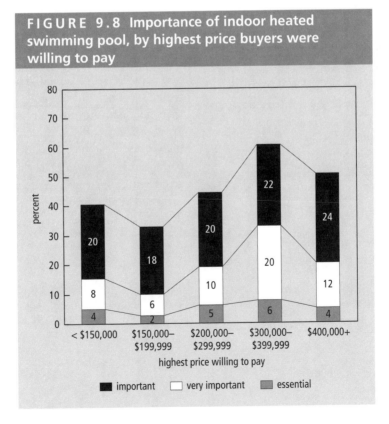

FIGURE 9.8 Importance of indoor heated swimming pool, by highest price buyers were willing to pay

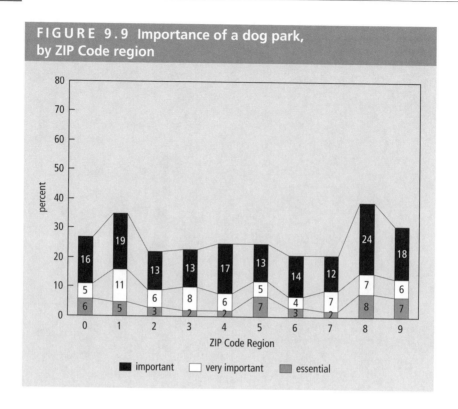

FIGURE 9.9 Importance of a dog park, by ZIP Code region

Dog Park

Compared with other areas, respondents in ZIP Code Regions 0, 1, 8, and 9 were most interested in having a dog park (fig. 9.9). Twenty-seven percent or more of them rated a dog park as important+. With the exception of the category of respondents seeking a $150,000–$199,999 home, some 30% of householders considered a dog park important+.

Staying Active and Engaged

Buyers who are 45+ want communities that offer plenty of opportunities for exercise and keeping

fit. Therefore, walking trails can be an important asset. Other on-site amenities can run the gamut from swimming pools to dog parks, but the appeal of each depends on individual market factors, including geographic region. Also keep in mind that more active community programming that gives residents off-site opportunities to get to know their neighbors can add value to lifestyle communities. For example, there is nothing like an adventure such as river rafting to foster friendships and bond a group together. Lifestyle is more than the on-site amenities.

10

Home Preferences: Footprint, Floor Plan, Style

Every home purchase is a series of trade-offs. Buyers may want a large home with plenty of yard, and they can have it, but the trade-off is being farther away from the center of town, or paying dearly for a lot closer in to the city.

For example, my husband and I live in the middle of the woods 11 miles south of town. Our house is lovely, but we only have two bedrooms, so we decided we would buy a second home in town. I insisted that our in-town home have three bedrooms—enough so that when we have company, everyone could be in a bedroom instead of sleeping on the Hide-a-Bed in the den. So, the search began. When it was completed I had my three bedrooms: two in the country and one in town. It was not exactly what I had planned, but we fell in love with a one-bedroom, one-bath house; its location (just two blocks from the historic square); and its beautiful, small, and manageable yard. Thank heavens there is a nice hotel a block away!

The Basic Home: Maximum Square Footage for the Money

Despite the common perception that everyone who is 55+ wants to downsize, the survey results demonstrate that, as with other age-groups, 55+ buyers want maximum square footage for their money. The data show that even householders planning to move to an active adult community aren't necessarily looking to downsize. Essentially equal percentages of 55+ householders that said they planned to move to an active adult community (34%), were considering both active adult and all-age communities (36%), or planned to move to an all-age community (36%) strongly agreed that they wanted the maximum square footage their money will buy (fig. 10.1). Generally, we like room to spread out, and 72% of middle-American householders ages 45+ agreed they wanted the largest basic home they could get for their money. Within that demographic, 37% of both the 45–54 and 75+ age-groups strongly agreed that they wanted the maximum size their money would buy.

At the same time, householders willing to pay more for their homes also were more willing to compromise on size. Forty-six percent of those planning to spend less than $150,000 for their homes wanted as much space as their money would buy. In contrast, 28% of respondents planning to spend $400,000 or more wanted as much space as their money would buy. Many boomers and silents who are able to invest more in their homes want better, but not necessarily bigger, homes.

Opinions on square footage varied somewhat by region, with the strongest preference for space among Midwesterners (ZIP Code Region 6). More than 80% of them strongly agreed or agreed that they wanted maximum square footage for the money (fig. 10.2).

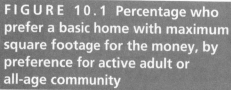

FIGURE 10.1 Percentage who prefer a basic home with maximum square footage for the money, by preference for active adult or all-age community

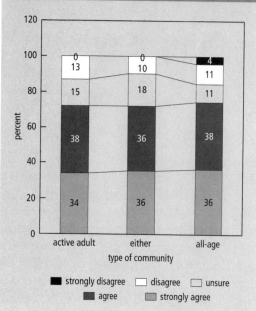

A Top-Quality Smaller Home

At the same time, the data show that buyers will compromise on size if that means they can upgrade the quality of their homes. Roughly half (53%) of middle-American householders ages 45+ who were planning to move agreed that they would prefer a smaller home, with everything top quality, to a larger home.

Moreover, the proportion that expressed a preference for a smaller home with everything top quality increased with increasing age. Although 14% of respondents ages 45–54 strongly agreed that they would sacrifice space if they could have a top-quality home, 25% of 75+ householders said likewise.

Preference for a smaller but top-quality home did not differ by the amount respondents planned to spend. The same proportion that planned to spend $400,000 as those planning to spend less than $150,000 were willing to make the trade-off of size for quality.

A majority of 55+ householders planning to move to an active adult community and those

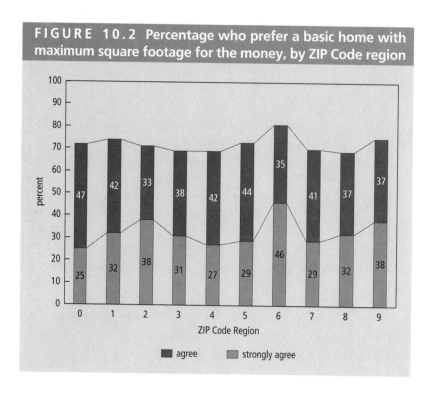

FIGURE 10.2 Percentage who prefer a basic home with maximum square footage for the money, by ZIP Code region

planning to locate in an all-age community were willing to trade size for quality as well. The preference for quality over size is more pronounced in the active adult group, however; more than two-thirds of them strongly agreed or agreed that they preferred a smaller home with everything top quality (fig. 10.3).

ZIP Code Region 8 had the highest proportion of respondents—more than two-thirds—that strongly agreed or agreed that they preferred quality over size. ZIP Code Region 2 had the smallest proportion (45%) willing to trade size for quality. Less than 50% of householders in other ZIP Code regions preferred quality over size as well (fig. 10.4).

Townhome or Condominium

Townhomes or condominiums were not a popular option among survey respondents; however, 23% were willing to make the trade-off from single-family to multifamily options to get

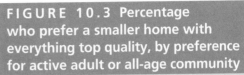

FIGURE 10.3 Percentage who prefer a smaller home with everything top quality, by preference for active adult or all-age community

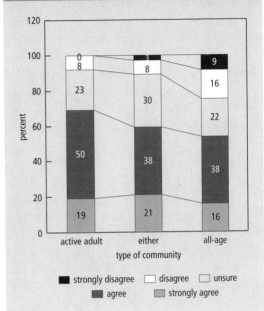

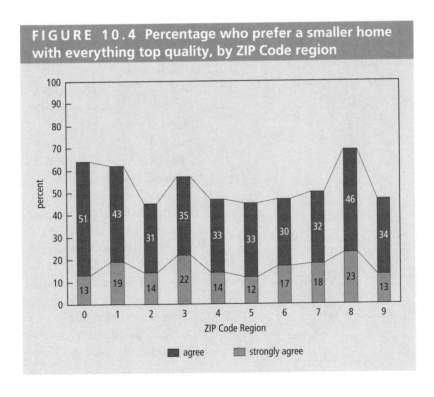

FIGURE 10.4 Percentage who prefer a smaller home with everything top quality, by ZIP Code region

a better home or a home in a better location (fig. 10.5).

The proportion willing to consider a condominium or townhome increased with the buyer's age. Roughly twice the percentage of 75+ buyers (35%) as buyers 45–54 (16%) agreed they would prefer a townhome or condominium instead of a single-family home if they could live in a better location or have upgraded home features.

The most pronounced difference of opinion was between those planning to move to an active adult community and those planning to go to an all-age community. Among those ages 55+ who were planning to move to an active adult community, 43% said they would trade a single-family detached home for a townhome or condominium if they could have nicer features or a better location for a lower price. Only 17% of 55+ householders planning to move to an all-age community were willing to make this trade-off.

A scant 11% of the households in the Mid-South (ZIP Code Region 7) would trade a single-family detached home for a townhome or condominium, while 31% of 45+ households in the Northeast (ZIP Code Region 0) would do so (fig. 10.6).

Spec, Custom-Built, or Existing Home

When asked to choose from among a brand-new home offered by a builder (spec home), a custom home built on their own lot, or an existing home, the largest proportion of 45+ middle-American householders (40%) was unsure which type of home they would purchase. Thirty-five percent preferred an existing home; 15% preferred to build a home on their own lot; and 10% would consider a new home already constructed by a builder.

Preferences for spec, existing, or custom homes vary somewhat by consumer age. Thirteen percent of householders ages 55+ said they would prefer a new home built by a builder compared with 8% of respondents ages 45–54. More respondents in the younger group preferred an existing (not new) or custom home (fig. 10.7).

Households that planned to spend less than $150,000 on their homes were most likely to purchase an existing home (56%), but 23% of those planning to spend $400,000 or more said they wanted a custom home built on their lot (fig. 10.8).

Among those seeking an active adult community, 28% planned to buy a new home offered by a builder, 24% planned to buy an existing home, and 42% were unsure of the type of home they would buy. Only 6% of these householders said they would build a custom home.

Of the buyers ages 55+ planning to move to an all-age community, only 8% said they would purchase a new home offered by a builder. Instead, 38% of these buyers planned to purchase an existing home, and another 40% were unsure of the type of home they would purchase (fig. 10.9).

The greatest proportions of householders that preferred an existing home, by region, were in New York and Pennsylvania (47%) and the Pacific Coast states (48%). Householders in the Western Mountain Region (ZIP Code Region 8)

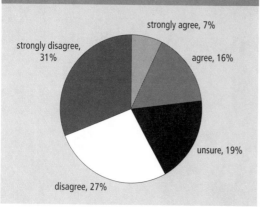

FIGURE 10.5 Percentage who would trade single-family home for multifamily home if it allowed them to move to a more desirable location or have nicer home features

strongly agree, 7%

agree, 16%

unsure, 19%

disagree, 27%

strongly disagree, 31%

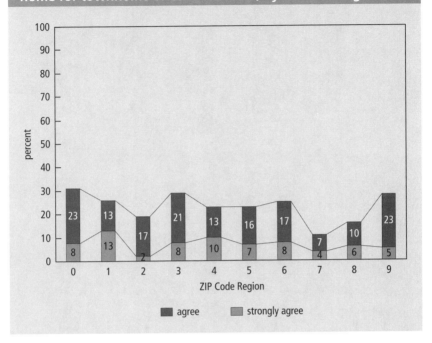

FIGURE 10.6 Percentage who would trade single-family home for townhome or condominium, by ZIP Code region

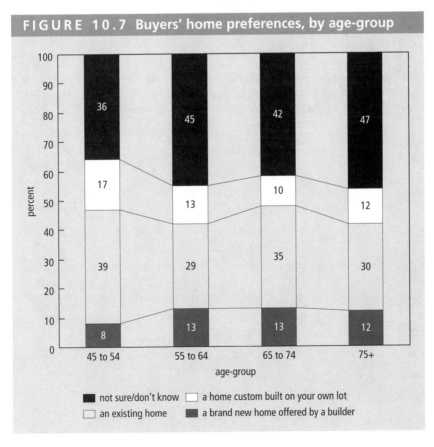

FIGURE 10.7 Buyers' home preferences, by age-group

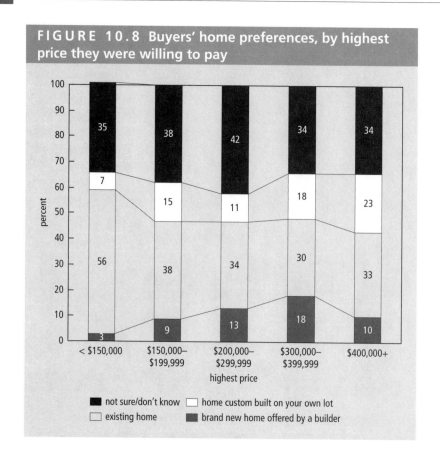

FIGURE 10.8 Buyers' home preferences, by highest price they were willing to pay

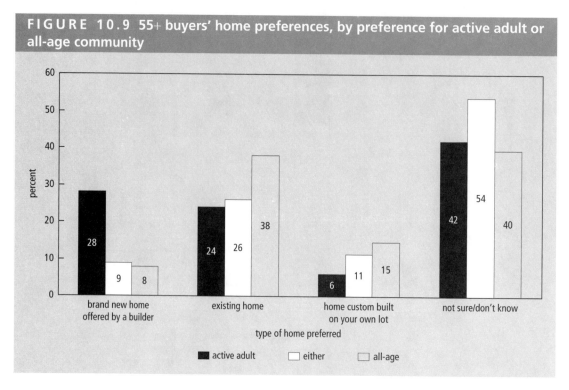

FIGURE 10.9 55+ buyers' home preferences, by preference for active adult or all-age community

were the most likely to want to purchase a new home offered by a builder. The smallest proportions interested in building a custom home on their own lot were in the Northeast, 3% in ZIP Code Region 0, and in the Midwest, 5% in ZIP Code Region 6 (fig. 10.10).

Floor Plan

Roughly half of middle-American householders ages 45+ planning to purchase a home within the next five years want a single-story, single-family detached home. Another 23% want a two-story detached home; 12% want an attached home; and 12% would consider purchasing a condominium in a multifamily, multistory complex.

The largest proportions of 45+ householders that planned to buy single-story, single-family detached homes either were unsure of whether they wanted a custom home, other new construction, or an existing home; or they believed they would

buy a new home (already constructed) from a builder. Respondents who preferred a two-story detached home said they were most likely to purchase an existing home (45%), but 29% of those who preferred a two-story detached home were unsure of whether they would buy a new or existing home, or have a custom home built for them. Twenty-two percent planned to have a custom home built.

Respondents who said they would purchase an attached home were divided equally between a preference for a new home offered by a builder (23%) and an existing home (23%). Forty-nine percent were not sure of whether they would purchase new construction, an existing home, or a custom home.

The proportion of householders that wanted a single-story detached residence increased from 51% among the 45–54 age-group to 59% among the 65–74 age-group. Conversely, the proportion that wanted a two-story detached home declined

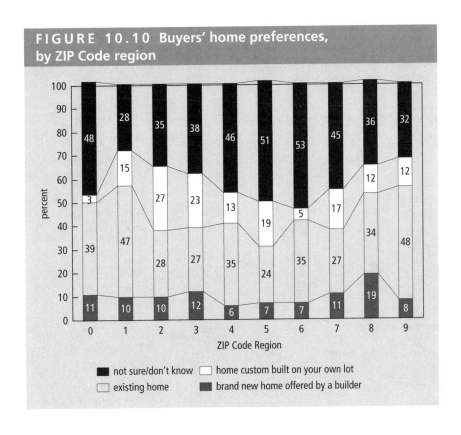

FIGURE 10.10 Buyers' home preferences, by ZIP Code region

with age—from 27% among those 45–54, to 11% of those ages 55–64, to roughly 5% of householders ages 65+. The proportion that wanted an attached home increased from 9% of householders ages 45–54 to 20% of those ages 65–74. Most of those who wanted to reside in a multifamily, multistory complex (29%) were 75+. Only 13% of the 45–54 age-group wanted to live in a multifamily, multistory residence.

Among 45+ middle-American householders, the proportion that wanted a single-story detached home decreased as desired home price increased. The proportion that wanted a two-story detached home increased with desired home price.

People 55+ who planned to move or who might move to an active adult community were more likely to consider an attached or multifamily residence than 55+ householders who planned to move to an all-age community (fig. 10.11).

The strongest preference for any type of housing that survey respondents were asked about was

in the Midwest (ZIP Code Region 6), where 69% of middle-Americans ages 45+ wanted a single-story detached residence. More than 60% of respondents in Region 5, the upper Midwest, and Region 7, which includes Texas, also preferred this option. The strongest preference for a two-story detached home was in the Mid-Atlantic (Region 2) and in New York and Pennsylvania, with 36% and 32% of respondents, respectively, preferring this option. The largest proportions of respondents interested in an attached home were in the Mid-South (ZIP Code Region 8), where 27% would consider an attached home, and in the Western Mountain region (ZIP Code Region 8), where 27% would consider that option (fig. 10.12).

Lot Size

Although the survey didn't ask specifically about lot or yard size, 20 years of other combined research on home buying habits has revealed that buyers are willing to exchange lot size for just about every other attribute of their homes and communities. In other words, lot size is the least important consideration for buyers. It has a major impact on their willingness to purchase a home only after they have decided on the style of community in which they want to live (e.g., urban, suburban, or rural). People planning to move to rural areas are more likely to want larger yards, whereas those moving to suburban communities are likely to accept homes with smaller lots. Some buyers insist on acreage; others are happy with a small courtyard.

Home Size

Although buyers understand the trade-offs with home size as well, more than half want at least 2,000 sq. ft. At the same time, most 45+ middle-American householders (75%) that plan to buy a home within the next five years wanted less than 2,500 sq. ft. The largest proportion (27%) wanted a home of 2,000–2,499 sq. ft. (fig. 10.13).

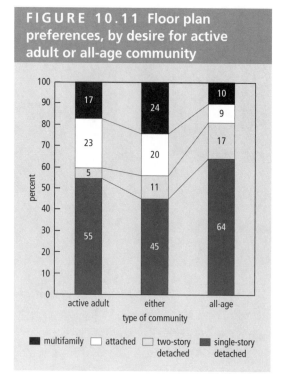

FIGURE 10.11 Floor plan preferences, by desire for active adult or all-age community

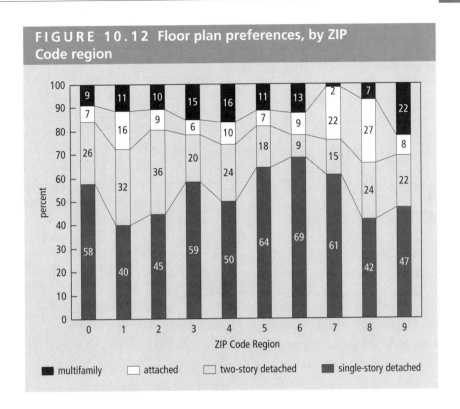

FIGURE 10.12 Floor plan preferences, by ZIP Code region

These preferences vary, however, by segment of the 45+ market. For example, the proportion of householders that preferred a 1,000–1,499 sq. ft. home tripled from 15% of those in the 45–54 age-group to 49% of those who were 75+. However, the proportion that wanted the smallest home—1,000 sq. ft.—did not increase with age. The proportion of householders that wanted a home of 1,500–1,999 sq. ft. increased from about 20% of 45- to 54-year-olds to roughly 30% of the other two age-groups. About 30% of 45- to 54-year-olds and 55- to 64-year-olds wanted 2,000–2,499 sq. ft.

Clearly, respondents understood the size/price trade-off in home buying. Consider that 66% of householders willing to pay less than $150,000 for their home wanted a home of 1,000–1,499 sq. ft., but only 4% of householders planning to spend $400,000 or more for a home wanted a home so modestly sized. In contrast, 35% of the latter group wanted 3,000 or more sq. ft.

Respondents also had differing size expectations based on whether they were planning to relocate to an active adult or an all-age community. Nearly double the percentage of respondents ages 55+ (33%) planning to move to an active adult community compared with those planning to go to an all-age community

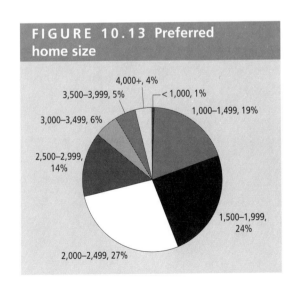

FIGURE 10.13 Preferred home size

wanted 1,000–1,499 sq. ft. Only 8% of those planning to move to an active adult community wanted a home of 3,500 sq. ft. or more, but an even smaller proportion—4%—of those planning to live in an all-age community wanted a home that large.

The majority of people living in a home of less than 1,500 sq. ft. wanted a home larger than their existing one. The largest proportion of people in homes between 1,500 and 2,499 sq. ft. wanted new homes the same size as their existing homes. A majority of those whose homes were 2,500 sq. ft. or larger wanted smaller homes. Note that the greatest proportion that wants to downsize from a home of between 1,500–2,999 sq. ft. wants a new home that is about 500 sq. ft. smaller. Those whose current homes are 3,000+ sq. ft. want to downsize to homes of about 2,000 sq. ft.

Number of Bedrooms and Bathrooms

One-bedroom residences are unpopular, even among single-person households. Fifty-one percent of middle-American householders 45+ who will live alone want 2 bedrooms, and 33% want a 3-bedroom home. Almost twice as many households planning to move to active adult communities compared with their counterparts in the same age-group who were planning to move to all-age communities wanted 2 bedrooms, but the majority of 2-person households (59%) wanted 3 bedrooms. The number of bathrooms desired loosely correlates with the number of bedrooms survey respondents said they wanted. Figure 10.14 shows the number of bedrooms desired by the highest price buyers were willing to pay. Figure 10.15 shows percentages of re-

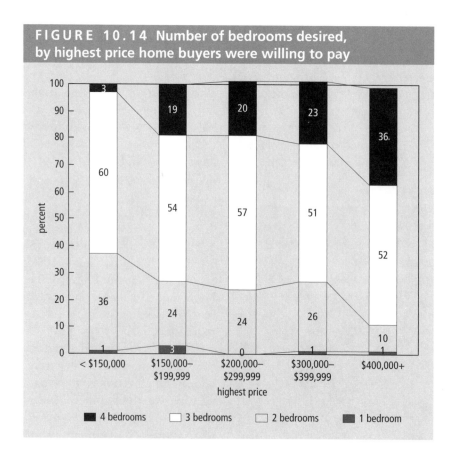

FIGURE 10.14 Number of bedrooms desired, by highest price home buyers were willing to pay

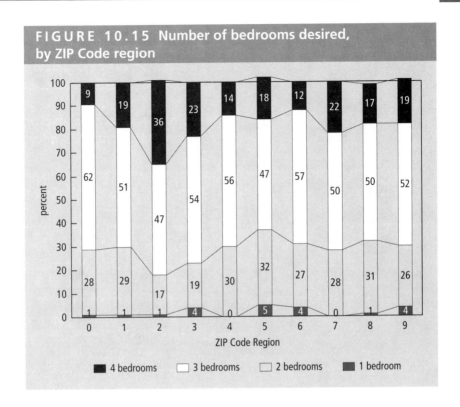

FIGURE 10.15 Number of bedrooms desired, by ZIP Code region

spondents in each ZIP Code region who wanted 1, 2, 3, or 4 bedrooms.

Garage Spaces

Builders and developers considering sacrificing garages for more density might want to think again. Only 4% of survey respondents said they did not want a garage, but three-quarters of 45+ middle-American householders said an enclosed garage is important and 48% said it is essential (fig. 10.16). Of the vast majority who wanted a garage, 57% wanted 2 garage spaces, 20% wanted 3, and 19% preferred 1 (fig. 10.17).

In addition, real-world buying behavior reveals that buyers will pay a little more for a home with a 2-car garage over a 1-car garage, or for a garage instead of a carport. This outcome holds true even for those who had said they would settle for 1 of the 2 lesser choices.

While it's certainly not an exciting feature, a two-car garage is a must. In the Northeast land is at a premium. Even in the light of this fact, with husband and wife either continuing to work or just the sheer fact of having two cars, this necessitates the inclusion of a two-car garage, even at the possible loss of density.

—**RONALD BONVIE**
President
Four Winds Developments
Mashpee, NH

Although the preferred number of garage spaces appears to decline with age, the importance of having a garage does not decline significantly. Moreover, although a somewhat larger proportion of respondents seeking active adult communities than those planning to move to

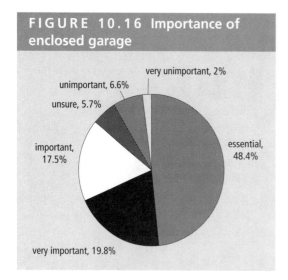

FIGURE 10.16 Importance of enclosed garage

very unimportant, 2%
unimportant, 6.6%
unsure, 5.7%
important, 17.5%
essential, 48.4%
very important, 19.8%

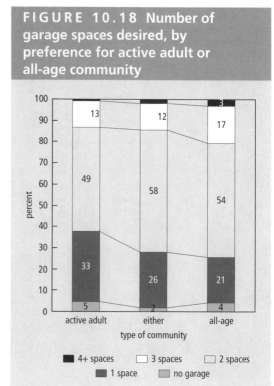

FIGURE 10.18 Number of garage spaces desired, by preference for active adult or all-age community

■ 4+ spaces □ 3 spaces □ 2 spaces
■ 1 space ▨ no garage

all-age communities opted for 1 garage space (fig. 10.18), more than 80% of both groups said an enclosed garage was important, very important, or essential.

At the same time, the number of desired spaces varies by preferred type of home (fig. 10.19). Householders planning to move to a 2-story detached home were the most likely to prefer 3 garage spaces (29%) or 4 or more garage spaces (11%). Differences according to the highest price buyers were willing to pay and regional prefer-

ences were evident as well (figs. 10.20 and 10.21).

Basements

To see genuine division in America, just ask people about their preference for a basement. Although 50% said a basement was unimportant or very unimportant, 40% of 45+ middle-American householders surveyed said a basement was important, very important, or essential (fig. 10.22). Although preferences are evident by region (fig. 10.23), as home buyers age and have more difficulty climbing stairs, they are willing to forego a basement. Even in tornado-prone areas where homes typically have had basements, there are good alternatives, such as a large closet in the center of the home. A basement was less important to respondents planning to move to an active adult community than for those planning to relocate in an all-age community (fig. 10.24).

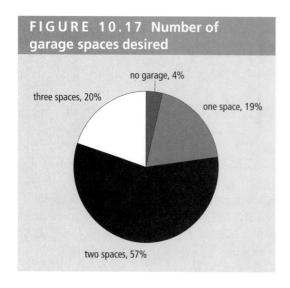

FIGURE 10.17 Number of garage spaces desired

no garage, 4%
three spaces, 20%
one space, 19%
two spaces, 57%

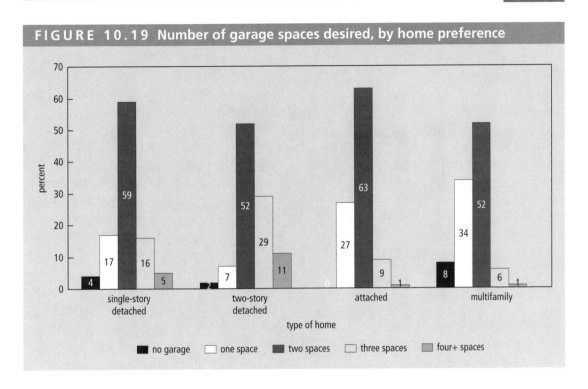

FIGURE 10.19 Number of garage spaces desired, by home preference

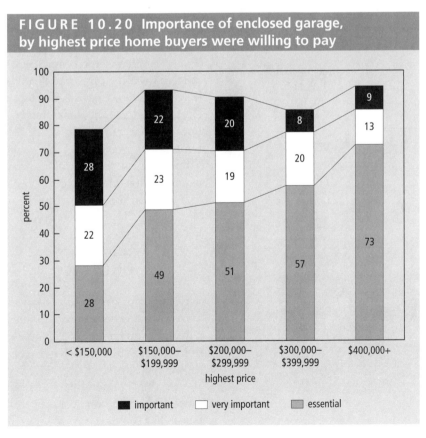

FIGURE 10.20 Importance of enclosed garage, by highest price home buyers were willing to pay

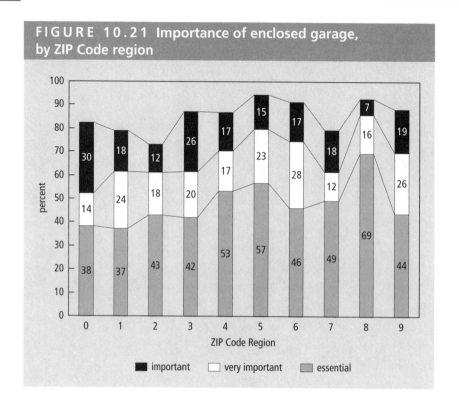

FIGURE 10.21 Importance of enclosed garage, by ZIP Code region

Universal Design

To gauge their interest in aging-in-place features, survey participants were asked to respond to the following statement: "Beautiful homes can be de-

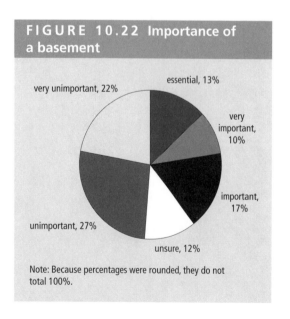

FIGURE 10.22 Importance of a basement

Note: Because percentages were rounded, they do not total 100%.

signed and built today that look like traditional homes, but that will accommodate changes of the residents over their lifetime. That is, regardless of how your abilities may change from an illness, accident or age, you should be able to live and function in your home without difficulty." Then respondents were asked, "How likely is it you would purchase this type of residence?" Forty-six percent of householders ages 45+ said they were very likely or likely to purchase a home with universal design features (fig. 10.25). Surprisingly, householders ages 55–64 were more likely to say they would purchase a universally designed home than were householders ages 75+ (fig. 10.26).

Respondents' likelihood of purchasing a home with universal design features increased somewhat with their desired price point and markedly with their preference for an active adult community (fig. 10.27). By region, respondents in the Mid-Atlantic (ZIP Code Region 2) and Western (ZIP Code Region 8) states were the most likely to purchase a universally designed home (fig. 10.28).

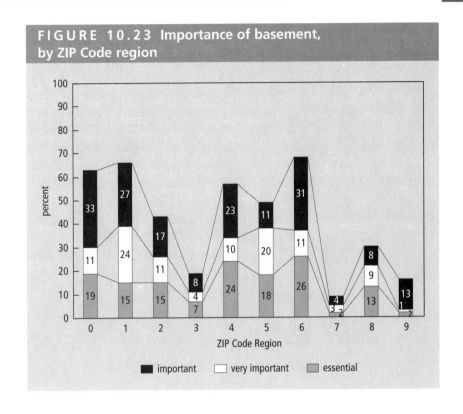

FIGURE 10.23 Importance of basement, by ZIP Code region

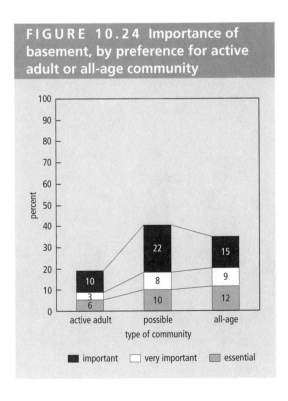

FIGURE 10.24 Importance of basement, by preference for active adult or all-age community

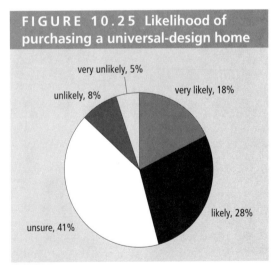

FIGURE 10.25 Likelihood of purchasing a universal-design home

very unlikely, 5%

very likely, 18%

unlikely, 8%

likely, 28%

unsure, 41%

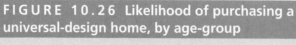

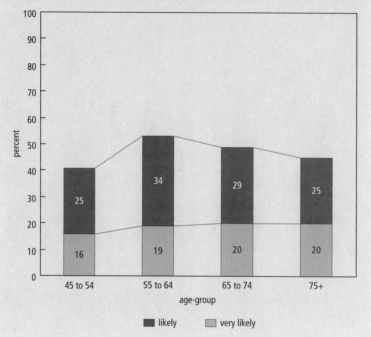

FIGURE 10.26 Likelihood of purchasing a universal-design home, by age-group

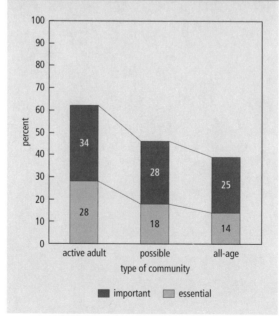

FIGURE 10.27 Likelihood of purchasing a universal-design home, by preference for active adult or all-age community

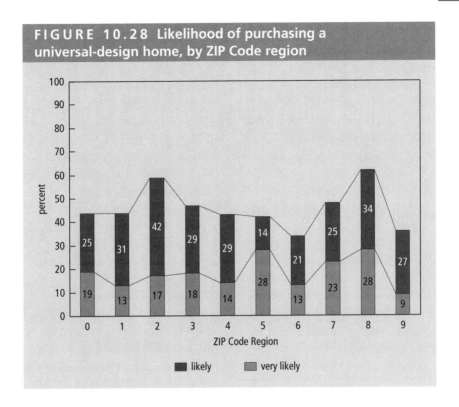

FIGURE 10.28 Likelihood of purchasing a universal-design home, by ZIP Code region

Market Insights

These survey results offer important insights for builders into the types of homes that will sell to 45+ buyers in their particular market—whether they are aiming their product at active adults or another market segment.

Generally, home buyers want maximum space for the dollar spent, but some—particularly those looking toward active adult communities—are willing to trade size for a better quality home. Moreover, people planning a move to an active adult community are more likely than those moving to an all-age community to be interested in a new home offered by a builder. Furthermore, 55+ households planning to move to an active adult community are more likely to purchase a townhome or condominium than the same age-group planning to move to an all-age community.

At the same time, not every maturing household plans to downsize, and the size of the homes that consumers will buy is influenced strongly by the size of their current homes. In fact, if they are in small homes today, they are more likely to want to upsize than downsize. Only a small proportion of respondents, even those in the silent and GI generations of 75+ home buyers, wanted a one-bedroom residence. Most respondents in all age-groups prefer three bedrooms, but many will trade a bedroom for another living space. However, home buyers moving to active adult communities are more likely to prefer two bedrooms than the same age-group moving to all-age communities.

Finally, a desire for a smaller vehicle doesn't translate into a desire for less garage space. Although a few householders are willing to accept a one-car garage, or even a carport, most prefer at least a two-car enclosed garage.

11

Rooms and Spaces in the Home

Mﾞore than 40 recent studies of 45+ consumers have revealed that despite differences in geographic region, economic circumstances, household composition, and overall housing likes and dislikes, 45+ home buyers as a group have similar preferences for their home's design.

This chapter summarizes results of these various studies to provide guidance to builders and developers about plans and options that might appeal to the lifestyles and tastes of boomers and silents in their own particular markets. Of course, no builder or developer should attempt to enter this or any market without conducting more detailed consumer research and feasibility studies. Chapter 12 offers suggestions for doing so as well as pitfalls to avoid.

> *The features that best sell the home start with the area around the community and its location; then, the buyer looks for a nice kitchen and owner's suite.*
>
> —**ALAN SPRINKLE**
> Senior Vice President Operations
> Regency Homes
> Des Moines, Iowa

The 'Ideal' Home

Although markets vary, the following are the "Top 10" attributes potential home buyers have listed most frequently when given a blank sheet of paper and asked to describe their ideal home:

1. large kitchen
2. open floor plan
3. attached garage with direct entry into the house
4. low or no maintenance
5. large closets
6. pleasing natural light
7. gas (preferred) or wood-burning fireplace
8. energy-efficient design, construction, appliances, systems, and features
9. large bedroom, bathroom, and closet as part of an owner's suite
10. outdoor living space, such as a porch, deck, balcony, or patio

Bedroom Preferences

An overwhelming majority—97%—of 45+ households say that privacy is one of a home's most important attributes. It's no surprise, then, that 90% of these households prefer that guest bedrooms be located on the opposite side of the house from the owner's suite. The owner's suite usually includes the largest bedroom, a walk-in closet, and a bathroom.

Sitting Areas

About half of prospective home buyers would prefer more space in public areas of the house to added seating in the owner's suite. Nevertheless, 25%–55% of buyers say seating in the owner's suite is important to them.

Dual Owner Suites

Most 45+ middle-American home buyers do not believe that dual owner's suites are important enough to spend additional money on them, although the proportion that preferred dual owner's suites increased with home value.

When included, these suites typically are on opposite sides of the home, or one is on the ground floor and the second is upstairs. Some are designed cleverly so that the bathroom in the second suite also serves as the guest bathroom.

A few buyers want three or four suites that include a bedroom-bathroom arrangement. Although that is a delightful convenience for those who entertain a number of overnight guests, most 45+ middle-American households are content with guest bedrooms close to the second full bath, but not necessarily arranged in a suite.

Closets

A walk-in closet in the owner's suite has become the standard in home design because consumers prefer it over the reach-in closet. Moreover, many 45+ consumers would prefer to have two walk-in closets in the owner's suite. So even though a walk-in closet may not provide any additional storage space, builders are wise to incorporate it. Where the luxury of two walk-in closets is not possible, the space can be separated to make it easier for partners to share one closet. Roughly half of households also would like a walk-in closet in the second bedroom.

Bathroom Features

Of the many available bathroom features, the greatest proportion of home buyers want a walk-in shower, linen closet, and exhaust fan (table 11.1).

Survey data also help answer one of the most frequently asked questions posed by builders and developers: "What bathing fixtures should be in the owner's suite?" (fig. 11.1)

Walk-in shower. A master bath should always include a walk-in shower—one with a threshold no higher than 4 in. (It does not need to be a roll-in shower, because most people who use wheelchairs are capable of getting into a shower with a

TABLE 11.1 Percent who said specific bathroom features are important

Room or Feature	Average	Variance (+/−)
Walk-in shower	92	10
Linen closet	90	10
Exhaust fan	89	5
Water temperature control	80	15
Ceramic tile	68	15
Private toilet compartment	60	15
Dressing/makeup area	59	20
Multiple shower heads	53	20
Deep soaking tub	49	20
Whirlpool tub	47	20
Heat lamp	35	20

FIGURE 11.1 Owner's suite bathroom bathing fixture preference

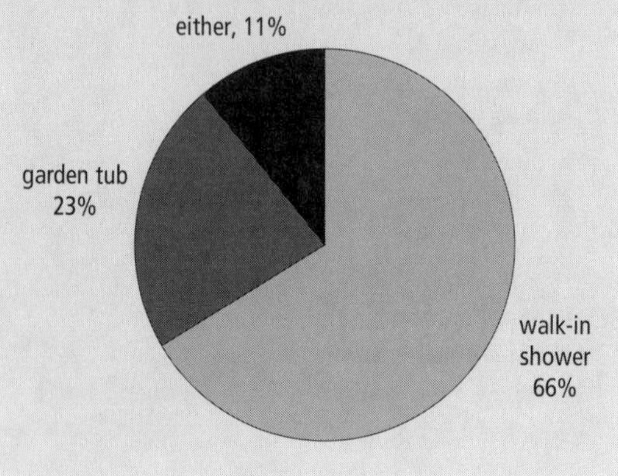

either, 11%

garden tub 23%

walk-in shower 66%

threshold.) The shower should include a seat with back, controls within reach, and an *attractive* (not hospital-like) grab bar or two. If the bath includes a garden tub (preferred by 23% of buyers), the tub also should have grab bars.

Two sinks. More than three-quarters of home buyers say that dual vanity sinks in the owner's suite bathroom are important, and roughly half say they are extremely important. Therefore, even when a home is built for a single person, dual sinks may increase a home's resale marketability.

Toilet chamber. Roughly 60% of the 45+ households like having a private toilet chamber in the owner's suite, and some 40% say a separate toilet chamber is extremely important.

> *Today we are including more home office space, more luxurious showers in a master bathroom without a tub, and a recharge station where you can set up chargers for phones, PDAs and laptops.*
>
> —KEN SIMON
> Chief Operating Officer
> Leewood Real Estate Group
> Staten Island, New York

Kitchens

Although Americans may not be cooking much, we're having more fun in the kitchen. It seems enigmatic in this age of heat-and-eat, prepackaged meals that kitchens usually contain the most frequently requested upgrades, including professional-grade stoves, Sub-zero refrigerators, and wine chillers. Multifunction entertainment centers also have entered the mix of kitchen must-haves for some consumers. Still, although media centers have become a popular upgrade in the 45+ middle-American kitchen, the most desirable features and upgrades remain the more traditional—a built-in microwave, pantry, and island (table 11.2). In order of preference, consumers want an island without any additional

TABLE 11.2 Percent who said specific kitchen features and upgrades are desirable for residence

Room or Feature	Average	Variance (+/−)
Built-in microwave	88	5
Pantry	86	5
Island	80	10
Drinking water filtration	73	20
Special storage for appliances	70	20
Corian	66	20
Cabinet upgrades	64	20
Island with cooktop	60	20
Upgraded appliances	60	20
Island with sink	55	20
Built-in desk	55	20
Trash compactor	48	20
Recycling center	47	20
Hot water dispenser	42	20

features, an island with a sink, or an island with a cooktop.

The trade-off with kitchen upgrades is not only cost, but the space they consume. With the kitchen being the center of the household, most buyers want a multifunction space for entertaining, dining, relaxing, and taking care of household business. These spaces must be attractive, functional, have efficient storage, serve multiple purposes, and be easy to maintain.

Beyond the Basic Rooms

Beyond the bedrooms, bathrooms, kitchens, and family rooms, buyers want additional spaces in their homes. For example, although some two-thirds of middle-American home buyers ages 45+ want a great room, one-third want something else. That something else may be a separate living room, den, dining area, or kitchen. Using data gathered from a number of studies, table 11.3 shows other desirable rooms and spaces from most- to least-preferred by consumers.

Storage

Storage is king. Approximately 80% of households in several studies of home buyers have concluded that buyers want a specific storage area in their home. In our survey, 55% of respondents said a large utility storage area attached to the house was important, very important, or essential.

Households will take the storage just about any way they can get it, but buyers typically prefer to have easily accessible storage at ground level. The most desirable locations include the garage or a storage room in the residence. To satisfy demand for storage, in lieu of basements some clever developers have started building small interior

TABLE 11.3 Percent who said additional rooms and spaces are desirable for residence

Room or Feature	Average	Variance (+/−)
Storage area	80	5
Laundry room	80	15
Deck, patio, porch	71	5
Enclosed porch	65	10
Great room	65	10
Sun room	61	20
Den, playroom, TV room	59	15
Breakfast area	58	5
Study or home office	52	5
Private garden space for each home	51	5
Dining room	46	20
Formal living room	44	20

rooms that can be used for storage or as hobby or computer rooms. In fact, many 45+ households prefer this storage area at ground level instead of a basement. If storage cannot be at ground level, other options include an attic above the garage or house, a second floor of the home, or a basement.

In short, storage should be placed anywhere it can be without eating up living space. Closets near the front door, broom closets in or near the kitchen, linen closets in the hallway and bathroom, extra storage space in the laundry or utility room, large-scale storage in the garage, a purpose-built storage room, a loft, or an attic will add functionality and value to a home.

Laundry Room

Although not as glamorous as an owner's suite, gourmet kitchen, or loft, a laundry room is an essential home feature. However, its location is not a given. Some consumers want the laundry room near the garage and/or kitchen, and some prefer to have it near the owner's suite. Only a small proportion of households want the laundry area in the garage or basement.

For a two-person or single-person household, a laundry area in the owner's suite makes sense. That's where most of the soiled laundry will come from and where it will have to be put away again after cleaning.

Outdoor Space

Living areas, even kitchens, have expanded into outdoor spaces, which are taking on a more finished appearance. Outdoor kitchens, fireplaces, dining areas, sound systems, and a deluge of furnishings and ornamentations draw the outdoors into the living space of a house. Slightly more than 70% of 45+ middle-American households thought that an outdoor living area (deck, patio, porch, screened-in porch, enclosed porch, or sunroom) was an important area of the home. Most households will want to include at least one outdoor living space in their home plans.

Sunroom

Whether it's called a three-season room, Florida room, Arizona room, or lanai, an enclosed patio, porch, or deck space is a popular upgrade. With plenty of sun, fresh air, and a view, it is often the most frequently used room in the house. Studies for dozens of developments have found that many households will pay a premium to have this space added as an upgrade.

Study or Home Office

With multiple personal computers, printers, and other paraphernalia, many home buyers are looking for places to use these tools and stow them attractively. Slightly more than half of 45+ households want a home office, and many want two. Homes without offices or other computer spaces often end up with the computer on the kitchen or dining room table. Well-designed spaces for using and storing equipment so that it is accessible, but not in the way, are a significant asset to a home.

Garden Space

About half of 45+ middle-American households enjoy gardening. Therefore, homes that incorporate flower beds around the house, in courtyards, and in outdoor living spaces will be attractive to many buyers. A mudroom with utility sink and storage for gardening supplies, then, is a sensible upgrade as well.

Dining Room

Dinner parties and entertaining at home are becoming more important. Therefore, dining spaces are an important feature for many 45+ households. About half of them want a formal dining area, whether in a separate room, an alcove, or as part of another space. This room is not just for family gatherings at holidays and birthdays, but for entertaining friends. It should

- be near the kitchen but separate enough to hide dirty dishes

- be expandable
- accommodate a dining table and space for a serving table or buffet
- incorporate variable lighting.

About half of the households that want a formal dining area also want other dining spaces in the home. About half want a formal dining room as well as a smaller informal dining area either in or outside of the kitchen. The other half prefers to have a dining room and an eat-in kitchen island.

Green Features

Builders and developers ask if home buyers want an energy-efficient home and if these consumers will pay for energy-efficient upgrades. The answer is not only yes, but a resounding affirmation of energy-efficient green homes and sustainable building.

Although typical consumers may not know how to create a more ecologically sustainable home and environment, they want to reduce environmental impact. Even if they are not ready for a straw-bale home, buyers who can afford to are more likely to purchase a home with upgrades and options that are beneficial to the planet than a home without them.

For example, all but a few studies completed in the past three years found that prospective home buyers were willing to pay for energy-efficient homes that would exceed Energy Star standards. In fact, several studies that tested buyers' willingness to pay $10 more per sq. ft. for energy-saving features found that a higher proportion of households were willing to purchase a home that exceeds Energy Star standards at the higher price than one that meets Energy Star standards at a lower price. Furthermore, 67% of prospective home buyers in one study said they would purchase a home with a heat recovery system for an additional $2,500, but 83% would pay an additional $12,000 to include a heat recovery system, insulated panel exterior walls, and R-50 insulation. As long as the home is within the price range

Energy Star

Energy Star is a program started in 1992 by the U.S. Environmental Protection Agency to reduce energy consumption and greenhouse gas emissions. The program has grown into a worldwide effort to promote energy-efficient home construction and consumer products. Energy-Star-rated new homes must have higher levels of insulation inspected for proper installation; complete framing and air barrier assemblies that enable insulation to perform at its full rated value; windows that meet or exceed Energy Star requirements; high-efficiency and properly sized heating and cooling equipment appropriate to the climate; and water heating, lighting, and appliances that are more energy efficient.

TABLE 11.4 Percentage range of 45+ middle-American housing consumers who want home systems

System	Range (%)
Energy management system	60–85
Multiple phone lines	50–75
Security system	50–75
Multi-zone HVAC	50–75
Whole house wiring high-speed internet	45–70
Lighting control system	45–70
Electronic air cleaner	33–60
Central music system	25–50
Central vacuum	25–50
Home entertainment system prewiring with plasma screen and speakers	25–50
Built-in entertainment center in family room	15–35
Intercom system	10–25

of the prospective home buyers, they are willing to pay more to upgrade the home to higher energy standards.

Home buyers also are willing to pay more to lower a home's environmental impact. From rainwater collection systems and gray water recycling to solar panels and natural landscaping, green building features increase market share—the proportion of consumers who say they will buy a home.

Home Automation Systems and Technology

The 45+ middle-American household loves technology as long as it is functional and easy to use; however, only a small percentage are technophiles eager to purchase and try every new gadget and gizmo on the market. Most 45+ middle-American home buyers know that technology is ever changing and becomes obsolete. Therefore, they are not inclined to want technology built into their homes. Still, they seek systems that will save money, improve efficiency, accommodate future technologies, add convenience, and entertain (table 11.4).

For example, home buyers want a simple method to control their energy consumption and reduce their energy bills. Therefore, the most frequently preferred technology upgrade is an energy management system. Some 60%–85% of survey participants have said they want such a system. Also, a majority of 45+ buyers would like to have a security system that is easy to arm and disarm.

Because consumers realize that entertainment-related systems will become obsolete, they are not as popular as the aforementioned options. Having lived through LPs, eight-track tapes, cassettes, CDs, DVDs, and now MP3 players and iPods, boomers, especially, know better than to expect a technology to last more than a few years.

Universal Design

Although they don't know precisely what it is, the 45+ middle-American household wants an accessible home. They depend on the builder to know how to provide a home that is chic, yet barrier free.

Most 45+ households are interested in step-less entries; low or no thresholds; wider doorways; single-story living or two-story homes that are designed to accommodate an elevator if needed in the future; and at least one walk-in shower. Buyers' knowledge of an accessible home typically ends there. As a result, many buyers then opt for 10 ft. ceilings, kitchen cabinets that reach to the sky, and light fixtures that require more than a step stool to change the bulbs.

As the boomer market becomes a greater force among homeowners, ever-increasing numbers of buyers will expect their living environments to better accommodate them. Demand for accessibility will affect every aspect of the house, including ingress and egress, moving throughout the residence, lighting (natural and artificial), bathroom fixtures, and maintenance and management of the systems and structure of the home. Consumers will expect a barrier-free home and maintenance-free features inside and out.

Ceiling Height

There is no hard-and-fast rule concerning ceiling height—at least for the country as a whole. Ceiling height preferences vary by region, price, and style of home buyers seek. Some 30%–50% prefer a 9 ft. ceiling, 10%–40% prefer 10 ft. ceilings, and 20%–30% want 8 ft. ceilings.

Home Features Boomers and Silents Want

In addition to the attributes most frequently cited for the ideal home, the features most desired by 45+ middle-American households include the following:

- owner's suite with a bedroom, bathroom, and walk-in closet
- guest bedrooms separated from the owner's suite
- storage, storage, and more storage
- walk-in shower in the owner's suite, and if space is available, a separate bathtub
- built-in microwave, pantry, and kitchen island
- kitchen upgrades including natural stone countertops, cabinets, water filtration, and media centers
- office space and/or other defined areas with storage that can function as office space
- laundry room on the ground floor (not in a basement or garage)
- outdoor living space
- home that will function over a person's lifetime
- systems that will save money, lower a home's impact on the environment, and improve occupants' ability to stay put as they age

Although these features are likely to attract most home buyers, they are only the beginning of that special recipe that will make your home and community a winning combination with buyers. Each and every market in the United States has its own nuances, character, and combination of factors that must be understood and tapped for a new community to be a robust success.

12

What Home Buyers Expect, Want, and Will Pay

This book has offered a comprehensive view of 45+ home buyers and what they want. However, builders and developers need to spend some time learning who their specific customers are, what they want, what they are willing to pay for, and how much they are willing to spend. Studies of thousands of households that form the foundation of this book reveal that the baby boom, silent, and GI generations are not monolithic groups but instead include a variety of market segments. Moreover, decades of work on hundreds of individual properties and developments in every state and in a few foreign countries reveal that builders don't have to go out of the country, across country, out of state, or even to a different city to find households with a variety of preferences. Even buyers purchasing in the same development and spending roughly the same amount of money can have vastly different desires. Therefore, a few weeks invested in understanding the similarities and differences among your particular customers will pay off in faster home sales, products priced appropriately (often at levels higher than a builder anticipated), and an apt match of buyers' desires to a particular community and home.

The next step is to conduct objective and critically timed research to uncover

- who your customers are
- what they want
- what will compel them to leave their current homes

- how much they will pay for the homes and lifestyle you have to offer

To assist builders and developers further in these efforts, this chapter

- reviews basic information about 45+ home buyers and renters
- offers suggestions for learning, and helping customers learn, what they want and how much they are willing to pay
- discusses how to ensure that planning, design, construction, positioning, marketing, and sales will be successful

45+ Home Buyers and Renters: No Hurry to Move

Home buyers ages 45+ usually aren't in a situation in which they have to move. Rather, they are at a point in life at which they would like to move to a place where they really want to live. They take time to think about the things they do and do not want; they talk with friends; and they size up what they want and need in a house and community. Some will have a list and even put together ideas for a floor plan, but most just have a mental picture, so they will look to the builder and developer to "wow" them with a community and home.

As with other market segments, referrals among 45+ buyers matter. Many will follow friends to a community. Still, with nothing forcing them to move, they do not have to tolerate inefficiency in sales, communication, or construction. Some buyers may be shopping for homes and communities as a recreational activity—researching particular areas of a state or the country, visiting communities and locations, and taking the time to discover the city, town, community, neighborhood, and home that fit them perfectly.

As these customers seek out new homes and communities, they look for builder expertise on universal design features, are willing to pay for certain upgrades, and understand the trade-offs inherent in obtaining both the home and lifestyle they desire. Because many 45+ households may be moving from full-time to part-time employment and have decreasing family responsibilities and smaller households, they can explore new options for living, spend money on upgrades instead of another bedroom, and pay for homes and communities that better represent who they are or who they want to become. Primary reasons 45+ buyers cite for moving are to have a new house, to live among others who are more like them, or to start over. Although real estate is part of the deal, the intangible fabric of the community is what really sells the home.

Builders and Developers as the Experts

Home buyers and renters want and expect expert knowledge and guidance from builders and developers. The key to customer delight is anticipating needs with product lines that buyers could not imagine on their own. Increasingly, this means incorporating, or at least being familiar with, universal design, home automation, and green building. As the boomer and silent generations purchase homes for what is an increasingly active phase of life, they will want them to accommodate new hobbies and interests, be easy to maintain, and facilitate their long-term needs.

Market Research Pitfalls

Builders sometimes assume they know what prospective home buyers and renters want simply because as professionals they've been in the market

a long time and have seen what sells. One method of developing new products is to look at what is selling best in other communities, tweak it a bit, and offer that. Another often-used method, but a bit more expensive, is to build a few models and see what sells the best. Neither of these is ideal. The first approach limits the potential to fine-tune a product so that it sells even better, and the second approach is inefficient and expensive.

Worse than both of these approaches, though, is asking the wrong people to evaluate proposed product lines. For example, builders eager to enter a particular market sometimes invite friends, relatives, and acquaintances to complete surveys or attend focus groups. Invariably, these friendly audiences say the builder or developer has a great concept that is needed in the area. What the builder doesn't ask and doesn't hear is, "It's a great idea for someone else, but I don't have any intention of moving there." Even riskier is to proceed on a project with virtually no input, as the following sad story illustrates.

A Cautionary Tale

An experienced builder new to the 50+ market decided to develop small active adult communities for people 55+ to compete with a seasoned active adult developer in town. He gathered dozens of floor plans from other communities and sorted them by type, basic floor plan patterns, wide or narrow lots, attached or detached, and other similarities. After weeks of painstakingly analyzing all of the differences and repeated elements and patterns of the homes, he came up with "ideal" floor plans and decided what his residential product would be.

The builder did not want to invest in feasibility studies; he already had demographic data collected to document that 55+ households lived in the area. He also didn't want to survey consumers because he had invested months studying floor plans. However, to convince the city council that the targeted households in the area wanted the community, he commissioned some quick telephone surveys to see what consumers thought about the concept. A majority thought the idea of the community described to them over the phone was appealing.

Most important, though, is what the builder did not find out. He did not

- have a market feasibility study completed or test the positioning and marketing of the community with target customers
- determine whether or not the locations were appealing to target customers or if the locations met any of the criteria important to the target market sector
- ask the customers what they wanted in their homes, streetscapes, and floor plans
- learn whether the targeted customers would be willing to pay the desired price for the homes and services he was planning

> *I have learned to ignore my personal likes and dislikes and make decisions based on what my target market is telling me. The most common mistake I see is builders/developers making product and marketing decisions based on their own personal choices, rather than what research is telling them the market will like. I like hearing, "the data says" rather than "I like. . . ."*
>
> —**KENNETH A. SIMONS**
> Leewood Real Estate Group
> Staten Island, New York

In short, the builder believed that walking through a few of the other communities in town and analyzing other communities' plans were sufficient to determine what consumers would buy. Therefore, he assembled land for three age-qualified communities, built clubhouses and sales offices, began the models, and initiated sales.

The builder had imparted his wisdom to a handpicked marketing team (none of whom was over 30), which positioned the communi-

ties and developed marketing collateral. "Caring relationship," "quality," "maintenance-free," "burden-less," "community activity director," and "concierge" were among the catchphrases used. The marketing team wanted to ensure that 55+ households knew their lives would be peaceful, tranquil, and bring no greater stress or excitement than the putting green.

The few homes that sold were purchased by people in their late 70s. The company, in its 80th year and fourth generation, went bankrupt and shut its doors.

A visit to two of the three sites and a mystery shop of the three sales offices prior to the bankruptcy offered clues to the lackluster customer reaction to these age-qualified communities. The communities were poorly located; the home designs were neither appealing nor accessible; and the sales centers and staffs were uninviting and unwelcoming. Specifically, one community was situated in an area with minimal drive-by traffic. The other had no drive-by traffic, was next to a railroad track, and was difficult to find. Although the communities were in urban areas, none of them were within walking distance of anything. The homes were small, with narrow porches and spindle-like wooden columns on the front of the house and a front entry without steps. They suggested a house for someone's great grandmother. Other entrances to the homes had steep stairways. Inside, small bathrooms and small kitchens that looked like an afterthought were accompanied by extra-wide doorways, but no railings on the steps to the basement.

When a mystery shopper visited the sales center, staff failed to introduce themselves or even ask the potential buyer's name. Once they did acknowledge the customer, they were eager to show floor plans, talk about home designs, and quickly close the sale—relating nothing about the community.

Lessons for Builders

That story of woe offers a half dozen lessons for other builders entering the active adult market.

1. Buyers are looking for more than a floor plan.
2. Home designs must seriously consider consumer's tastes and desires.
3. The community, particularly its location, and not just the home, will significantly influence a purchase decision.
4. A "caring," "peaceful," "tranquil," and "burden-less" community will not attract the majority of the 45+ market looking for a vibrant, fulfilling lifestyle.
5. Builders and developers should ask, rather than tell, their target market who they are and what they want.
6. Although a community may be age qualified, that is not, nor should it be, its major selling point. Research points toward some common features 45+ middle-Americans seek in a home; however, their experiences, interests, upbringing, careers, partners, and hundreds of other individual characteristics—not their age—define who they are.

Ask First, Build Later

The best way to learn what consumers want is to ask them. The key question builders must ask their potential market is, "Will you move to this community in this location and buy one of these houses at this price within the next year?" Fortunately, objective research that asks this question of potential buyers usually can be completed for less than the net profit of one home. Moreover, valid, reliable research usually increases profitability, yields happy customers, and generates qualified leads and referral sales. "I love the research results, but what I really love are the buyers it generates and the fact that they tell their friends," says one builder.

The size, expense, and risk of a building venture dictate the type and amount of information to be gathered. Without exception, though, market research should

- define the target customers for a home or community

- learn what they want and are willing to purchase
- find the price they are willing to pay
- discover the price elasticity

Where fewer than a dozen homes are likely to be sold in a year, one-on-one interviews, focus groups, and written surveys are reasonable research methods. Properties requiring significantly more investment call for more extensive research: market feasibility studies; focus groups; telephone and written surveys; research seminars; and, when possible, conjoint analysis.

Conjoint analysis allows research subjects to consider several key attributes of location, home, and community simultaneously, so it closely resembles the purchase experience. Therefore, it is the best method to learn what people want, what they're willing to pay, and what's most important to them in deciding to purchase. Using conjoint analysis, builders can tap into potential buyers' reactions to all of the variables customers deem important in purchasing a home. When builders put that knowledge to work in choosing a location, then design homes and communities accordingly, they increase the likelihood that buyers will choose their homes over a competitor's.

Ultimately, research must help builders answer this question: If we build a house and community for a particular price, position, and lifestyle in a certain way and begin marketing and sales by X date, how many sales per month can we anticipate?

Research Guidelines

1. Focus on households considering a move.
2. List the specific information needed from the research, decisions that the research will influence, who will use it, and when it is needed.
3. Ensure objectivity by avoiding relatives, friends, and acquaintances as research subjects.
4. Target research to specific objectives.

12 Critical Questions to Answer

1. What are the household characteristics of potential buyers in the target market?
2. How many of these households exist?
3. What proportion of these households are likely to purchase a home within the year?
4. How likely are they to purchase the homes at the particular location and for the builder's desired price?
5. What sizes, styles, floor plans, and level of quality does the target market sector want?
6. What amenities are important, and how much is the target market willing to pay to have the desired amenities?
7. What services does the target market sector want and how much are the people in that sector willing to pay for them?
8. What messages and images appeal to the target customers and yield a higher proportion of customers saying they will buy a home in the community?
9. What lifestyle or positioning of the community appeals to the target market?
10. What is it about the lifestyle and positioning that is appealing to them?
11. What attributes are desired for lots, views, landscaping, and streetscapes in the community?
12. What architectural design will have curb appeal for the targeted market, and what will consumers be willing pay for it?

Sales Offices: An Essential Source of Market Information

Market research doesn't, and shouldn't, end with a community's groundbreaking. To ensure that a community continues to appeal to buyers, builders must gather and review information about their customers systematically. Ask the sales staff to use a brief, standard information form and to politely beg people to fill it in. This form should ask the customer's year of birth, marital status, household composition, ZIP Code, how they learned about your community (give them a list to choose from), type of home or types of homes they are considering, approximate size of home they want, estimated value of their current home, and their desired price range for a new home. The information gathered helps to pinpoint where prospects live (the market area), who they are, and the marketing message or other element that attracted them to your sales office.

During their first visit to a sales office many people are reluctant to provide their addresses, telephone numbers, and e-mail addresses because they don't want to be pursued. Go ahead and accept their first name and allow them to omit any identifying information; but ask them politely to give you the rest of the information about themselves. Guarantee them that if they do not wish to be contacted, they will not be.

Also using a standard form, ask customers to rate various aspects of your floor and site plans on a scale of 1 to 10, in which a "1" represents absolutely not what they want and "10" is a perfect match. By going over these forms with customers, sales staff can probe further into what attracts customers and what does not. For example, a salesperson might inquire, "I see you really didn't like our floor plans with the courtyard porches on the side of the house instead of in the back; why?" or "You rated the Liberty house plan as an 8; what would make it a 10?"

By using standard forms to guide conversations with customers, builders can gather valuable information that can be fed back into their design, marketing, and sales efforts.

Post-Purchase Customer Satisfaction

Once a sale is made, customers should be surveyed about their purchase experience, home, and community. The goal is not just to receive a high satisfaction score that can be used in future marketing but rather to learn how to serve customers even better in the future. What grabs customers is not an abstract score but the tangibles that make a community look and feel great, a personable and genuine sales staff, an atmosphere that suits their lifestyle, and a lovely home at a price they were happy to pay. Ask buyers to rate every aspect of your community, homes, marketing, and sales center using the following scale:

- needs no improvement
- needs minimal improvement
- needs some improvement
- needs considerable improvement

Review the results, and use the information to inform decisions about future homes and communities.

Afterwords

Readers who picked up this book looking for a recipe or a cookbook for successfully building homes for boomers and silents may be disappointed or frustrated. Indeed, no formula exists that will appeal to buyers who are in diverse locations and from many age-groups and income levels, and who, despite having age in common, may be in various stages of life's journey. On the basis of ProMatura's extensive consumer research of those ages 45 and up, the soundest advice I can offer to builders and developers with regard to age is to take this factor out of the equation when determining what to build, where to build, and in what price range. Focusing on the age of a target market risks missing the multitude of other attributes and factors that are more likely than age to predict what the market prefers and is likely to purchase. Instead, focus on the specifics of the households who are interested in the particular product, place, and price you have to offer—whether you are selling a primary residence, second or vacation home, or retirement property. You can match your product to the specific home, community, and lifestyle preferences of your target market sector so that it will sell faster, at an optimal price, and make your buyers happier with their purchase.

With that said, I urge all builders and developers to pay attention to what potential home buyers have conveyed in our research. Their views will allow you to come close, in fact very close, to creating the homes and communities your target market sectors want. Use this information, then, to fine-tune your own market research about desirable neighborhood and community features as well as services, amenities, and programming that would appeal to buyers in your particular locale. If you ask objectively, systematically, and often, and believe what your customers are telling you, you can build the right house, in the right place, at the right time.

U.S. Zip Code Regions

0	Connecticut (CT), Massachusetts (MA), Maine (ME), New Hampshire (NH), New Jersey (NJ), Rhode Island (RI), Vermont (VT)
1	Delaware (DE), New York (NY), Pennsylvania (PA)
2	District of Columbia (DC), Maryland (MD), North Carolina (NC), South Carolina (SC), Virginia (VA), West Virginia (WV)
3	Alabama (AL), Florida (FL), Georgia (GA), Mississippi (MS), Tennessee (TN)
4	Indiana (IN), Kentucky (KY), Michigan (MI), Ohio (OH)
5	Iowa (IA), Minnesota (MN), Montana (MT), North Dakota (ND), South Dakota (SD), Wisconsin (WI)
6	Illinois (IL), Kansas (KS), Missouri (MO), Nebraska (NE)
7	Arkansas (AR), Louisiana (LA), Oklahoma (OK), Texas (TX)
8	Arizona (AZ), Colorado (CO), Idaho (ID), New Mexico (NM), Nevada (NV), Utah (UT), Wyoming (WY)
9	Alaska (AK), California (CA), Hawaii (HI), Oregon (OR), Washington (WA)

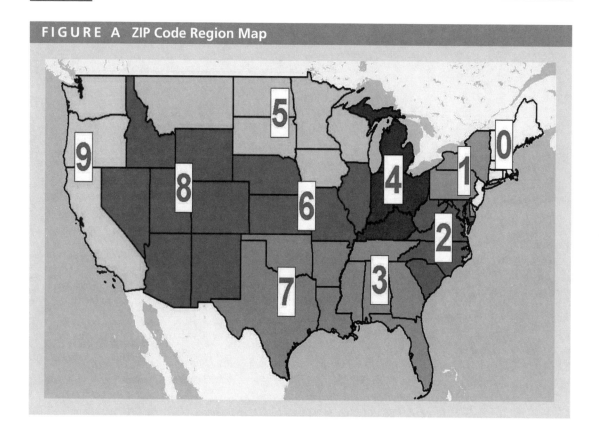

FIGURE A   ZIP Code Region Map

U.S. Census Regions

Four groupings of states (Northeast, South, Midwest, and West) were established by the U.S. Census Bureau in 1942 for the presentation of census data. They are as follows:

- Northeast Region: Maine, New Hampshire, Vermont, Massachusetts, Connecticut, Rhode Island, New Jersey, New York, Pennsylvania
- South Region: Maryland, Delaware, West Virginia, Virginia, Kentucky, Tennessee, North Carolina, South Carolina, Georgia, Florida, Alabama, Mississippi, Arkansas, Louisiana, Oklahoma, Texas
- Midwest Region: North Dakota, South Dakota, Nebraska, Kansas, Missouri, Iowa, Minnesota, Wisconsin, Illinois, Michigan, Indiana, Ohio
- West Region: Washington, Idaho, Montana, Wyoming, Oregon, California, Nevada, Utah, Colorado, Arizona, New Mexico, Alaska, Hawaii

Glossary

active adult community. A community for buyers or renters of a minimum age, usually 55–62. It may include single-family homes, townhomes, or condominiums. Although children are welcomed as guests, even for extended stays, they are not permanent residents. Amenities may include a clubhouse, swimming pool, tennis courts, walking trails, or other facilities or features. Active adult communities usually do not provide meals or housekeeping services. Residents generally have chosen them for

- access to amenities such as golf or swimming
- the opportunity to live near other people with similar interests and schedules
- reduced lawn and home maintenance
- the ability to "lock and leave" their homes
- community covenants that govern the community's character and appearance

55+ rental community. An apartment building solely for people who are 55 or older. It usually has some attractive amenities such as a card or game room, a central meeting area for residents, and a fitness center. Some have swimming pools, tennis courts, covered parking, and on-site beauty shops. Usually, services such as a dining program or housekeeping are not provided by the property management company nor are they included in the monthly fee.

continuing care or life care retirement community. A campus or complex that may have homes, condominiums, and/or apartment-style residences. A variety of services are usually available that may include meals, housekeeping, transportation, and social, educational, and recreational programs. The campus typically has care available for people who may need additional help with daily activities, such as meals, bathing, dressing, mobility, and managing medications, either temporarily or permanently. The campus usually includes a nursing care center. People who choose to move to these communities do so for the reassurance of lifelong care.

independent living community. A multifamily apartment complex that has an on-site dining program. Individual apartments usually have kitchens, but the community may have one or more dining rooms where most residents typically eat their largest meal of the day. Basic services such as housekeeping, transportation, and social, educational, or recreational programs may or may not be provided. Fees or rent are paid monthly.

assisted living community. An apartment-like building that provides housing and services designed to assist people with everyday activities. It provides three meals per day. Services usually are planned for each person based on his or her requirements but may include help getting from one place to another, bathing, dressing, and managing medications. Fees are paid on a monthly basis.

References

Joint Center for Housing Studies of Harvard University, Graduate School of Design, John F. Kennedy School of Government, State of the Nation's Housing 2006, Harvard University, Cambridge, MA, 2006.

Roberts, Sam; Ariel Sabar, Brenda Goodman, and Maureen Balleza Contributed Reporting. 51% of Women Are Now Living Without Spouse. *The New York Times,* January 16, 2007.

Simmons, T. and J. L. Dye, Grandparents Living with Grandchildren: 2000, Census 2000 Brief, US Census Bureau, October 2003.

U.S. Census Bureau, 2005 American Community Survey, Table R1101. Percent of Households That Are Married-Couple Families: 2005a.

———, Current Population Survey, America's Families and Living Arrangements: 2005, Table FG2, Married Couple Family Groups, by Family Income, Labor Force Status of Both Spouses, and Race and Hispanic Origin of the Reference Person: 2005b.

———, Current Population Survey, Table 2B. Presence of a Computer and the Internet at Home for People 18 Years and Over, by Selected Characteristics: October 2003, Internet Release, October 27, 2005: 2005c.

Index